PUZZLES FOR PLEASURE

BOOKS BY E. R. EMMET

THINKING CLEARLY

LEARNING TO THINK

LEARNING TO PHILOSOPHISE

BRAIN PUZZLER'S DELIGHT (PUBLISHED IN ENGLAND AS "101 BRAIN PUZZLERS")

PUZZLES

FOR

PLEASURE

E. R. Emmet

EMERSON BOOKS, INC.
Buchanan, New York

"TO MY WIFE —
*who provides the pleasure,
while I produce the puzzles.*"

Third Printing, 1977

COPYRIGHT © EMERSON BOOKS, INC. 1972
ALL RIGHTS RESERVED
STANDARD BOOK NUMBER 87523-178-0
LIBRARY OF CONGRESS CATALOG CARD NUMBER
71-189618
MANUFACTURED IN THE UNITED STATES OF AMERICA

Contents

Preface ix

PART I: ASSORTED—SIMPLE

1. Dames for Delphi 3
2. Election Speculation 4
3. Hilarious Holidays 5
4. The Temple of Torpor 6
5. Happy Birthdays 7
6. Nurses Relax 8
7. Mugs, Wumps and Others 9
8. Early Examination Exits 10
9. Quietly Home to Bed 11
10. The Black and White Foreheads 12
11. Ditheringspoon's Spanish 13

PART II: OUR FACTORY

12. I Sit Aloof 17
13. Just the Jobs 18
14. Body and Soul 19
15. Who Works When? 20
16. Chagrin for Charlie 21
17. When Are We? 22
18. Working to Rules 23
19. Who Lives Where? 24
20. Sammy the Soothsayer 25
21. Muscles Are Not Enough 26
22. Wealth, Happiness and Health 27
23. Mr. Bottles and Mr. Doors Go to the Convention 29
24. How Faithful Our Bottle-Washer 30
25. We Polish Up Our Percentages 31
26. How Vocal Our Voters 33
27. The Pay Roll Rules 35
28. The Higher the Truer 36

29. Who Are We?	38
30. The Sad Demise of Derrick Demmit	40

PART III: HOCKEY

31. 3 Teams	45
32. 5 Teams	45
33. 6 Teams	46
34. D Was Dumb	47
35. A Long Leg	48
36. Uncle Bungle Has His Moment	49
37. Four Figures False	50
38. Uncle Bungle Predicts	51
39. Some Ham Humor	53
40. Uncle Bungle Baffled	54
41. A Muddy Muddle	55
42. No Light of Truth	56
43. Zanies, Yodellers and X's	57

PART IV: CROSS NUMBER PUZZLES

44. Straight as a Die	61
45. Straight as a Cork-Screw	62
46. Uncle Bungle's Cross Number	63
47. A Three-Dimensional Cross Number	64
48. A Treble Cross Number	66

PART V: ASSORTED—NOT SO SIMPLE

49. Top Hats and Tails	71
50. The Older the Better	72
51. Orders about Orders	73
52. Who's Sitting Where?	73
53. In-Laws	74
54. Only Two True	74
55. Some Buffers Bite	75
56. The Island of Perfection	76
57. Round the Table	78
58. The Family Tree of Truth	79
59. The Dowells Do Better	80

60. Logic about the Logic Prize	81
61. Some Eccentric Sportsmen	81
62. Happy Family Unions	82

PART VI: SILENCE OR DECEIT—SOME MISSING AND SOME INCORRECT DIGITS

63. Multiplication and Division	85
64. 9 Digits Divided by 2 Digits	86
65. 7 Digits Divided by 2 Digits	87
66. 5 Digits Divided by 3 Digits	88
67. An Ascending Quotient	89
68. An Unconvincing Division	90
69. 4 Digits Divided by 2 Digits (all incorrect)	91
70. 6 Digits Divided by 2 Digits (all incorrect)	92
71. A Mistake in the Mistakes	93
72. Addition and Subtraction	94

PART VII: THE ISLAND OF IMPERFECTION

73. Unable, Unhappy and Unwanted	97
74. The Silent C	97
75. The Charming Chieftainesses	98
76. Ringing Is Not All Roses	99
77. Logic Lane	100
78. The Top Teams on the Imperfect Island	101
79. Some Tribal Wedlocks	102
80. The Immortal Wolla	103
81. Imperfect Wives and Daughters	104
82. Competitions in Imperfection	105
83. Uncle Bungle on the Island	106
84. The Tribal Trousers	107
85. Some Imperfect Wages and Taxes	108
86. The Years Pass	109
87. A Mixed Marriage	110
88. Uncle Bungle Fails Again	111
89. World Cup for the Wotta-Woppas?	113
90. Some Assorted Tribal Soccer	114

91. Our Factory on the Island	115
92. More Imperfect Soccer	117

PART VIII: ASSORTED—HARDER

93. The Age of Remembrance	119
94. Fast or Slow	120
95. Help-the-Boys Hall	121
96. The Light of Truth	122
97. Anstruther and Others	122
98. Abracadabra Avenue	123
99. What Tulsa Tortoise	124
100. The Wumbling Widgets	125
101. 13th Avenue	126
102. Refined by Mogrification	127

Solutions   129–310

Preface

The problems or puzzles of real life will not be solved here, and it is in fact a "Zany" world indeed in which my characters live. But for those who are interested in straight, logical thinking it has the merit that, unlike the real world in which we find ourselves, if we read the instructions carefully, however reminiscent they may be of Alice in Wonderland, and apply the simple principles of straight thinking, we shall arrive at a precise answer to the question that has been posed. Uncle Bungle may make his mistakes, but his slips will show unmistakeably. Except in the few cases where there is a precise statement to the contrary, uncertainties are out. We know where we are going, or rather we know where we are trying to go, and the gong of success sounds loud and clear when we have arrived. But for greater satisfaction in solving these problems the reader is advised to satisfy himself (or herself, and I disagree strongly with the view that is sometimes held that ladies are less logical) that the solution is correct and unique before looking it up.

Within each section the problems are arranged in what seemed to me to be approximately an ascending order of difficulty, though clearly there is room for some difference of opinion about this. And it is not easy sometimes to distinguish between harder, in the sense of having to take steps in which more subtle thinking is required, and harder in the sense of having to take a greater number of steps, each of which may be very simple.

Dr. Watson often expressed his amazement at the deductions of Sherlock Holmes, an amazement which, when the chain of thinking was explained to him, was replaced by a scornful description of the reasoning as obvious and easy. The difficulty lies, of course, not in understanding the successive steps in the argument, but in knowing which steps to take, and as in all processes of thinking this will depend very much on experience and on keeping the mind active and flexible.

I would make no great claims about success in the solving of these problems leading to more effective thinking in other fields, to promotion in business or politics, or the amassing of fortunes on the stock market. Their main purpose is to provide fun for those who enjoy this sort of thing, and perhaps for those who may be persuaded to try it. But nevertheless they certainly provide exercise for the machine which nature has given us for thinking, and provided that we clearly understand the essential difference between the closed system thinking of puzzles, and the open system of real life, and make due allowance for them, our better exercised and therefore more powerful minds should be more effective in all the fields of their endeavor.

Some of these problems appeared in my "The Use of Reason," and in "Learning to Think," both published by Longmans, and one of them was used as a "Brain Teaser" in the British "Sunday Times." But more than half of them have not previously appeared in print.

A mistake or a misprint in a book of this kind may obviously matter very much. All the problems and their solutions have been checked and re-checked by many friends. I mention particularly the late R. V. H. Roseveare, a very great mathematician, and a critical and therefore particularly helpful checker, also Richard Fletcher-Cooke, James La Trobe-Bateman, Robin Aizlewood, Richard Snedden, Thomas Livingstone-Learmonth, and John Gaskell who, between them, discovered a number of errors and made many valuable suggestions. It would perhaps be optimistic to hope that all errors have been nailed. If there are any still remaining there can be no doubt where the fault lies; we must blame Uncle Bungle. The typing of this book, especially of the solutions, required great skill, patience and accuracy. For all these virtues, and for an enormous amount of hard work I am most grateful to Mrs. J. H. Preston and Mrs. M. G. Whalley.

E. R. Emmet

PART I

Assorted — Simple

1 — 11

1. Dames for Delphi

Alpha, Beta, Gamma and Omega were four young ladies of ancient Greece who were training to become oracles; in fact only one of them did and she got a post at Delphi. Of the other three one became a professional dancer, another a lady-in-waiting and the third a harp player.

During their training, when they were practicing predictions one day, Alpha forecast that whatever else Beta did she would never become a professional dancer; Beta forecast that Gamma would end up as the Delphic oracle; Gamma forecast that Omega would not become a harp player; and Omega predicted that she would marry a man called Artaxerxes. The only prediction that was in fact correct was the one made by the lady who became the Delphic oracle.

Who became what? Did Omega marry Artaxerxes?

2. Election Speculation

On the Island of Perfection there are four political parties — the Free Food, the Pay Later, the Perfect Parity and the Greater Glory. Smith, Brown and Jones were speculating about which of them would win the forthcoming election.

Smith thought it would be either the Free Food Party or the Pay Later Party. Brown felt quite confident that it would certainly not be the Free Food Party. And Jones expressed the opinion that neither the Pay Later Party nor the Greater Glory Party stood a chance. Only one of them was right.

Which party won the election?

3. Hilarious Holidays

The natives of the island of Hilary (Hilarians) fix their feast-day (Whupie) according to the following rule:

"Whupie shall be on that Saturday which is as many Saturdays after the first Saturday in February as the last digit in the year or the last digit but one in the year whichever is greater (e.g. in 1963 it was 6 Saturdays after the first Saturday in February), unless it is a leap year in which case it shall be one Saturday later than it would be in a normal year."

(i) *When was Whupie in 1972? (March 1, 1972, falls on Wednesday)* (ii) *What was the last year in which Whupie was in February?*

4. The Temple of Torpor

If a man is over forty he cannot be a member of Yonkers Youngsters. Everyone of importance in Salem is either a member of Yonkers Youngsters or of Iowa Idlers or both. Only if one is allowed to enter the Taunton Temple of Torpor can one be a member of Iowa Idlers. You cannot both not be under thirty and be allowed to enter the Taunton Temple of Torpor.

Smith is someone of importance in Salem and is not under thirty. What else can you say about him?

5. Happy Birthdays

Alpha, Beta, Gamma, Delta and Epsilon have their birthdays on consecutive days, but not necessarily in that order.

Alpha's birthday is as many days before Gamma's as Beta's is after Epsilon's. Delta is two days older than Epsilon. Gamma's birthday this year is on a Wednesday.

On what days of the week are the birthdays of the other four this year?

6. Nurses Relax

Seven nurses A, B, C, D, E, F, G have a day off every week, no two of them on the same day.

You are told that A's day off is the day after C's; that D's day off is three days after the day before E's; that B's day off is three days before G's; that F's day off is half way between B's and C's and is on a Thursday.

Which day does each nurse have off?

7. Mugs, Wumps and Others

(Two statements are said to be contradictory if they cannot both be true and cannot both be false; two statements are said to be contrary if they cannot both be true but can both be false.)

The following statements are made, all about the same person:

(1) He is a Mug (3) He is a Blink
(2) He is a Wump (4) He is a Blonk

You are told that (1) and (2) are contradictories, that (2) and (3) are contraries and that (1) and (4) are contraries. What can you say about (3) and (4)?

Is it true to say that (i) all, (ii) some, or (iii) no Mugs are (a) Blinks (b) Blonks?

8. Early Examination Exits

Smith understood Mr. Jones to say that the earliest time for leaving the examination room was 2 hours before lunch or $1\frac{1}{2}$ hours after the examination started, whichever was the earlier. Robinson however was under the impression that Mr. Jones had said that the earliest time was $2\frac{1}{4}$ hours before lunch or 1 hour after the examination started, whichever was the later.

Smith and Robinson both left the examination room at the earliest possible moment, according to what they thought their instructions were. They both left at 11:15 A.M.

What can you say about the time of lunch and of the start of the examination?

9. Quietly Home to Bed

A notice goes up:

'All those who have not already received their subsistence allowance this week should report to the Deputy Assistant Bursar (D.A.B.) at 9 A.M. next Monday unless they failed to get it last week in which case they shall be deemed to be without visible means of subsistence and should not report to anybody anywhere unless they are over 21 and/or have a maternal grandmother who is still alive in which case they should either report to the Substantive Acting Registrar at 10 P.M. on Tuesday or to the D.A.B. at 12 noon on Wednesday according to whether their surname begins with a letter in the first half of the alphabet or not unless their paternal grandmother does not have or did not have a passport in which case they should report to the Warden as soon as possible unless they have already passed their driving test and/or do not have a bicycle in which case they should take a couple of aspirins and go quietly home to bed.'

Jones, aged 20, has not drawn his subsistence allowance this week or last. He has a bicycle but has not passed his driving test. Both his grandmothers are still alive and have passports.

What should he do?

10. The Black and White Foreheads

There are five men, A, B, C, D, E, each of whom has a white or a black disc attached to his forehead. Each man can see the discs worn by the other four, but is unable to see his own. If a man is wearing a white disc, any statement he makes is true; if he is wearing a black disc, any statement he makes is false. Statements are made as follows:

    A says: 'I see three white discs and one black.'
    B says: 'I see four black discs.'
    C says: 'I see one white disc and three black discs.'
    E says: 'I see four white discs.'

Deduce the colors of the discs worn by all five men.

11. Ditheringspoon's Spanish

Adams, Baines, Clarke, Ditheringspoon and Elliott are the only entrants for the school Spanish prize. After the result has been announced they speak as follows:

ADAMS: I was not fourth.
BAINES: I was one place lower than Clarke.
CLARKE: I won the Russian prize.
DITHERINGSPOON: I was one place higher than Adams.
ELLIOTT: I was three places higher than Clarke.

All the above remarks are true except the one that was made by the person who came last for the Spanish prize.

What was the order of merit for the Spanish prize? Can you say anything about the winner of the Russian prize?

PART II

Our Factory

12 — 30

12. I Sit Aloof

When I first employed Alf, Bert, Charlie, Duggie and Ernie in my factory, they used to sit for their many discussions round a circular table to emphasize that in my eyes they were all equal, though naturally as the Managing Director I sat aloof. In order to make it clear that there was no alphabetical superiority I insisted that no two men, the initial letters of whose names were next to each other in the alphabet, should be next to each other at the table. Charlie had Ernie's brother sitting on his right.

Where is everyone sitting relative to Charlie?

13. Just the Jobs

I find it interesting to look back to the days when I first opened my factory with Alf, Bert, Charlie, Duggie and Ernie, and had to decide how to allot the jobs of Door-Keeper, Door-Knob-Polisher, Bottle-Washer, Welfare Officer and Worker.

Not surprisingly they all had their opinions about it. Alf was most emphatic that whoever was to polish the door-knob, it should not be Ernie. Charlie was very keen to wash the bottles, but Ernie thought that Duggie was the most suitable person for this job. Bert hoped that Alf would be the worker and expressed a strong disinclination himself to be the Welfare Officer. Duggie warmly recommended Bert for the job of Door-Keeper.

After the jobs had been allotted it was interesting, looking back, to see that the hopes and recommendations of the Bottle-Washer and the Worker had not in fact been fulfilled, but that those of the Door-Keeper and the Door-Knob-Polisher had.

Who had which job?

14. Body and Soul

I have been trying to persuade my employees at Our Factory of the merits, whatever Shakespeare's Julius Caesar may have said, of an honest soul in a slim body, and the connection between weight and truth. Alf, Bert, Charlie and Duggie have become particularly weight conscious and they were making some remarks one day about the latest news which the scales had given them.

They spoke as follows:

ALF: Bert is lighter than Duggie.
BERT: Alf is heavier than Charlie.
CHARLIE: I am heavier than Duggie.
DUGGIE: Charlie is heavier than Bert.

I found it interesting to note that my theory was supported. In fact the only one of these remarks to be true was that which was made by the lightest of the 4 men (their weights were all different).

Arrange Alf, Bert, Charlie and Duggie in order of weight.

15. Who Works When?

When we first opened Our Factory it was obviously important to have clear rules about who was to work when.

I arranged that Alf, Charlie and Ernie should be present and Bert and Duggie absent on the opening day, and in order to regulate the attendance on all subsequent days, I laid down the following rules.

Alf will work if and only if Bert was present and Charlie absent on the previous working day.

Bert will work if and only if Charlie was present and Duggie absent on the previous working day.

Charlie will work if and only if Duggie was present and Ernie absent on the previous working day.

Duggie will work if and only if Ernie was present and Alf absent on the previous working day.

Ernie will work if and only if Alf was present and Bert absent on the previous working day.

Who was present on the factory's 100th day, 383rd day?

16. Chagrin for Charlie

No one knows better than I do what a competitive world we live in and I encourage my workers to take every opportunity of building up and exercising their competitive spirit. Charlie has got his very fully developed and it was with great chagrin that he told me one morning that he had not been first in a race that he had been having with Alf, Bert, Duggie and Ernie.

Duggie, he also informed me, was two places below Ernie, who was not second, and Alf was neither first nor last. I heard later from Bert that he was one place below Charlie.

Find the order in which they came in the race (no ties).

17. When Are We?

Alf, Bert, Charlie, Duggie, Ernie, Fred and George are having an argument about which day of the week it is.

They speak as follows:

ALF: The day after tomorrow is Wednesday.

BERT: No, it's Wednesday today.

CHARLIE: You're quite wrong; it's Wednesday tomorrow.

DUGGIE: Nonsense, today is neither Monday, Tuesday nor Wednesday.

ERNIE: I'm quite sure yesterday was Thursday.

FRED: No, you've got it the wrong way round. Tomorrow is Thursday.

GEORGE: Anyway yesterday was not Saturday.

Only one of these remarks is true. Which one? What day of the week is it?

18. Working to Rules

At different times and under different circumstances it has seemed to me to be desirable, bearing in mind especially the complications of human relationships, to have rules in Our Factory that were tailor-made for the particular situation.

The rules at one time were as follows:
1. If Alf is present Bert must be absent unless Ernie is absent, in which case Bert must be present and Charlie must be absent.
2. Alf and Charlie may not be present together or absent together.
3. If Ernie is present Duggie must be absent.
4. If Bert is absent Ernie must be present unless Charlie is present, in which case Ernie must be absent and Duggie must be present.

Production, for obvious reasons, must be on the basis of a seven day week; the public needs us. But ruts must be avoided and I felt it was essential that there should be a different set of workers present on each of the seven days.

What were the seven different sets of those present and absent, while the above rules were in force?

19. Who Lives Where?

I have been very fortunate in being able to make arrangements for all my staff, Alf, Bert, Charlie, Duggie, Ernie, Fred and George, as well as my Chauffeur, Hubert, and my secretary, Judith, to live in houses (all different, of course) on Domum Road. This road has houses numbered from 1 to 55. My employees all made statements about the numbers of their houses as follows:

Alf said that his number was 23 more than Bert's;
Bert " " " " " 16 less " Charlie's;
Charlie " " " " " 19 less " Duggie's;
Duggie " " " " " 12 more " Ernie's;
Ernie " " " " " 30 more " Fred's;
Fred " " " " " 17 less " George's;
George " " " " " 37 less " Hubert's;
Hubert " " " " " 12 more " Judith's;
Judith " " her " " 10 more " Alf's.

I discovered afterwards that one of these statements was not true.

Find the numbers of the houses in which all my nine employees live.

20. Sammy the Soothsayer

The employees of our factory have recently been taking an interest in foretelling the future and Alf, Bert, Charlie and Duggie persuaded Sammy, a local soothsayer, to work with them one evening.

These four had, not necessarily respectively, the jobs of Bottle-Washer, Door-Opener, Door-Shutter and Worker and I had told them that these jobs were shortly to be distributed between them in a different way so as to give them all a change. Each of my employees has a personal number which is naturally not known to anyone else. (These are all different positive whole numbers.)

In the course of the evening's session 13 statements or predictions were made, of which 10 were true and 3 false. I think that my workers would prefer that I should not say who made the false remarks, and I therefore set them out, in no particular order, as follows:

1. The future Worker's number is odd.
2. Charlie is the Bottle-Washer.
3. The present Bottle-Washer's number is 11.
4. The future Door-Shutter's number is even.
5. Bert is the future Door-Opener.
6. The future Worker's number is even.
7. The next Bottle-Washer will be Charlie.
8. Bert's number is 15.
9. Duggie is the future Door-Opener.
10. The future Worker's number is not 6 less than the future Bottle-Washer's number.
11. The future Door-Opener's number is 27.
12. Charlie's number is 11.
13. The present Worker's number is greater by 1 than the sum of the other 3 numbers.

Find the present jobs and the future jobs of Alf, Bert, Charlie and Duggie. What can you say about their personal numbers?

21. Muscles Are Not Enough

Workers in a modern factory need mathematical techniques as well as manual skills and so I recently gave Alf, Bert, Charlie, Duggie and Ernie a searching test on these lines.

The table below shows the answers given by each of them to five questions which I set, and also some particulars of the total number of points gained by each person and for each question.

No. of question	1	2	3	4	5	Total for each person
Alf	3	20°	17	20%	$15	30
Bert	7	10°	21	5%	$1	15
Charlie	4	18° 30'	45	5%	$143	30
Duggie	3	20°	63	10%	$4	
Ernie	4	12°	21	10%	$17	10
Total for each question	30	10	35	15	15	

If the answer was right I gave 10 points. If the answer was wrong I gave either 0 or 5 points according to my assessment of the validity of their method and the legibility of their writing. (If 2 of them got the same wrong answer it was possible for one of them to get 5 and the other 0.)

At least one person got each question right. What are the right answers, and how many points did each of them get for each question?

22. Wealth, Happiness and Health

On the rare occasions when the workers in my factory fall ill they do so unanimously, intensively and luxuriously. The precise nature of the highly infectious disease with which Alf, Bert, Charlie, Duggie and Ernie were recently afflicted need not concern us now. It resulted in them all being admitted to the Muddlestreet Hospital at the local seaside resort where we had been spending the day on our factory outing. As is customary in this splendid place they each had a private room and a private nurse.

The names of the nurses were Priscilla, Queenie, Rachel, Sarah and Tess; and their rooms were all on the tenth floor which has rooms numbered from 97 to 103 inclusive.

To preserve their alphabetical prestige the nurses all insist that a higher place in the alphabet gives them a claim to look after a lower-numbered room. (So that Priscilla will look after a room which has a number smaller than those looked after by the other 4, and so on.) Our patients are conscious from time to time and when they are they make some rather inconsequential remarks.

'Ernie has got Tess for his nurse', said Alf, 'lucky man!'

Bert seemed to be bubbling with knowledge. 'Sarah is Charlie's nurse', he said, 'and serve him right'. 'I can feel it in my bones', he went on, 'that Duggie's room number is 102, but that Alf's room number is not greater than 100'.

'Duggie's room number is a prime number', said Charlie, waking up suddenly and slipping as swiftly back into a deep torpor.

'Alf', said Duggie, 'is being looked after by Sarah'. 'What has he done to deserve that?', he inquired ambiguously.

Ernie was our mathematician even in slumber. 'A multiple of 7', he said quietly with his eyes shut, 'is what Bert's number is'.

An analysis of these statements in the light of the facts as I knew them showed that the remarks of only two of them were true — the two who were in the rooms with the highest numbers.

All the remarks made by the other three were false.

What was the name of the nurse and the number of the room of each of our five patients?

23. Mr. Bottles and Mr. Doors go to the Convention

We have recently been having a re-shuffle of jobs in Our Factory and there is still some uncertainty in the minds of my workers — Alf, Bert, Charlie, Duggie, Ernie, Fred and George — as to who is doing which of the seven basic jobs — Door-Opener, Door-Shutter, Door-Knob-Polisher, Bottle-Washer, Sweeper-Upper, Welfare Officer and Worker.

4 of them had been selected to represent the factory in a convention of other forward looking firms which was to discuss policies for the next 10 years. I had decided to call them Mr. Welfare, Mr. Sweepers, Mr. Bottles and Mr. Doors, but although each man knew the title which he himself was to wear he did not know who was to wear the others.

I traveled to the convention with our 4 representatives and made notes of various remarks which they made — as follows:

MR. WELFARE: 1. Fred is the Bottle-Washer.
                      2. Bert is the Worker.
                      3. Duggie is not Mr. Bottles.
MR. SWEEPERS: 1. Alf is the Worker.
                      2. Charlie is not Mr. Bottles.
MR. BOTTLES: 1. Ernie is the Welfare Officer.
                      2. Bert is the Bottle-Washer.
MR. DOORS: 1. Duggie is the Worker.
                      2. Charlie is the Bottle-Washer.
                      3. George does not have one of the door jobs.

It was interesting, but not perhaps surprising, that each statement in which the name mentioned was one of those present was true. Each statement, however, in which the name mentioned was one of the three who were not present was false. (No one made a statement mentioning his own name. The titles for the convention are not necessarily connected with their present jobs.)

Find the names and the jobs of our 4 representatives.

24. How Faithful Our Bottle-Washer

The 7 employees in Our Factory — Alf, Bert, Charlie, Duggie, Ernie, Fred and George — are all at work spasmodically. The Factory has full working days on Monday, Tuesday, Thursday and Friday and a half working day on Wednesday. The Factory's rules are as follows:
  (i) The Bottle-Washer must always be present.
  (ii) At least 2 employees must be absent each day.
  (iii) Bert must never be absent on the day after Alf is present unless Duggie is absent too, in which case both Bert and George must be absent.

Those present more than $2\frac{1}{2}$ days always tell the truth.

Those present less than $2\frac{1}{2}$ days never tell the truth.

Those present exactly $2\frac{1}{2}$ days make statements which are alternately true and false.

(Wednesday is the only day on which employees may work half a day.)

They make statements as follows:

ALF: 1. Duggie and Ernie are equally truthful.
2. George is the Bottle-Washer.
BERT: 1. I wasn't there on Wednesday.
2. George is never absent when Ernie is.
CHARLIE: 1. Bert and Duggie are equally truthful.
2. Bert was there on Wednesday.
DUGGIE: 1. Fred was only absent on Wednesday.
2. I was there for less than $2\frac{1}{2}$ days.
ERNIE: 1. Alf and I are never present together or absent together.
2. I wasn't there on Tuesday or Wednesday.
FRED: 1. Ernie always tells the truth.
2. Duggie was there on Monday and Tuesday.
GEORGE: 1. I wasn't there on Wednesday.
2. Duggie always tells the truth.

Who was present when? Who is the Bottle-Washer?

25. We Polish up Our Percentages

Alf, Bert, Charlie, Duggie, Ernie, Fred and George have been doing some hard thinking about the nature of their occupation and the amount of their pay, with special reference to their percentage differentials. Their jobs are, not necessarily respectively, and not necessarily distributed in the same way as at any other time, Door-Opener, Door-Shutter, Door-Knob-Polisher, Bottle-Washer, Sweeper-Upper, Welfare Officer and Worker. The Door-Knob-Polisher and the Sweeper-Upper never tell the truth; the Door-Opener and the Door-Shutter make statements which are alternately true and false or false and true; the other three always tell the truth.

Their wages per week are now and always have been an exact number of dollars: no one gets more than 400 dollars per month.

They make statements as follows:

ALF:
1. I get 20% more than Bert.
2. Bert gets 5% more than Ernie.

BERT:
1. Fred is the Welfare Officer.
2. I am not the Door-Opener.

CHARLIE:
1. George is the Bottle-Washer.
2. The Door-Knob-Polisher gets 15 dollars more than the Worker who gets 20% more than the Sweeper-Upper.

DUGGIE:
1. I get 10 dollars more than the Welfare Officer who gets $33\frac{1}{3}$% more than he used to.
2. Alf gets more than George.

ERNIE:
1. Alf had an 8% raise this week.
2. The Door-Opener gets 20% less than the Worker.

FRED:
1. Charlie is the Worker.
2. The Door-Shutter is the best paid of us all.

GEORGE: 1. I am not the Worker.
2. Bert is the Sweeper-Upper.
3. Ernie gets 5 dollars less than the Door-Knob-Polisher who gets 10 dollars less than Fred.

Find the wages and occupation of each man.

26. How Vocal Our Voters

There has been a fever of excitement at Our Factory about the forthcoming election.

As Managing Director I thought it would be interesting to ask Alf, Bert, Charlie, Duggie, Ernie, Fred and George how they would vote. They made some remarks about their intentions and were unable to resist the opportunity of saying something also about their fellow workers.

The reader will not be surprised to hear that not everything that was said was true. An analysis showed the following interesting facts.

Every statement made by a Socialist was false.

Every statement made by a Progressive was true.

The statements made by the Republicans were alternately true and false, in that order.

The statements made by the Democrats were alternately false and true, in that order.

(The jobs of my seven workers are, in no particular order, Door-Knob-Polisher, Door-Opener, Door-Shutter, Sweeper-Upper, Welfare Officer, Bottle-Washer and Worker.)

Their statements were as follows:

ALF:
1. I shall not vote Socialist or Republican.
2. None of us belong to the same party as Bert.

BERT:
1. I shall not vote Democrat.
2. George is a Progressive.

CHARLIE:
1. I shall vote Socialist.
2. Duggie belongs to the same party as the Bottle-Washer.
3. The Welfare Officer is not a Socialist.

DUGGIE:
1. I shall vote Progressive.
2. Bert is a Democrat.

ERNIE:
1. I shall vote Democrat.
2. George's party is different from the Door-Opener's.

     3. Alf is the Door-Knob-Polisher.
FRED: 1. I shall not vote Republican.
     2. The Sweeper-Upper is a Democrat.
     3. Alf and the Worker belong to different parties.
GEORGE: 1. I shall vote Republican.
     2. Duggie is a Socialist.
Find their jobs and political parties.

(*It may be assumed that if, for example, the statement* 'Alf belongs to the same party as the Worker' *is known to be true, Alf is* not *the Worker*.)

27. The Pay Roll Rules

In the early days of Our Factory there were only five employees — Alf, Bert, Charlie, Duggie and Ernie. Their jobs were, not necessarily respectively, Door-Keeper, Door-Knob-Polisher, Bottle-Washer, Welfare Officer and Worker.

The question arose as to whether, and if so what, they should be paid. It was clearly very difficult to arrange this in such a way as to reward merit accurately and avoid friction, but eventually, after much thought, I, as Managing Director, put up the following set of rules:

PAY

1. Alf is to get more than Duggie.
2. Ernie is to get 12 per cent more than the Bottle-Washer will when he receives the 10 per cent raise that he will be getting next month.
3. The Door-Knob-Polisher is to get 30 per cent more than he used to.
4. Charlie is to get 12 dollars less than 20 per cent more than the Welfare Officer.
5. No one is to get less than 200 dollars or more than 600 dollars a month.
6. The Door-Keeper is to get 5 per cent more than he would if he got 10 per cent less than Bert.
7. Everyone heretofore, now and hereinafter, has received, does receive and will receive, an exact number of dollars a month.

What are the various jobs of my employees and what monthly wage is each of them to get?

28. The Higher the Truer

Alf, Bert, Charlie, Duggie, Ernie, Fred and George, have been smitten with a mysterious disease. They have all been admitted to a hospital and have been given the private rooms numbered 98–104, which are consecutively situated round the central circular hall.

Those of them whose temperatures are under 99° never tell the truth unless they are wearing pink pajamas, in which case they make statements which are alternately true and false.

Those of them whose temperatures are over 102° always tell the truth.

Those of them whose temperatures are between 99° and 102° inclusive make statements which are alternately true and false.

Two of them are wearing pink pajamas, and the colors of the pajamas of the other five are flame, blue, green, yellow and aquamarine.

(Anyone who is known to make statements which are alternately true and false may start in the statements below with one which is either.) They make statements as follows:

ALF:
1. Fred's pajamas are aquamarine.
2. Charlie's room number is 2 greater than his temperature.
3. Duggie's room number is over 100.

BERT:
1. Alf's room number differs by 3 from Fred's temperature.
2. Duggie's temperature is 100°.
3. George's temperature is 98°.

CHARLIE:
1. Fred's temperature is 98°.
2. Ernie's room number is even.
3. Fred is wearing blue pajamas.
4. I am not wearing yellow pajamas.

DUGGIE:
1. Alf's pajamas are green.

　　　　　　2. My temperature is greater than my room number.
　　　　　　3. Alf's temperature is one less than his room number.
　　　　　　4. My pajamas are flame.
ERNIE:　　1. My pajamas are pink.
　　　　　　2. My temperature is under 99°.
　　　　　　3. Bert's temperature is 2° higher than Fred's.
FRED:　　　1. My temperature is not 101°.
　　　　　　2. Duggie's room number is a prime.
　　　　　　3. Charlie's temperature is over 102°.
　　　　　　4. Bert's and Ernie's rooms are next door to each other.
GEORGE:　1. My pajamas are not pink.
　　　　　　2. Charlie's temperature is 97°.
　　　　　　3. Charlie's temperature is over 102°.
　　　　　　4. The number of Fred's room is 2 greater than the number of Bert's.

Find for each of them his room number, the color of his pajamas, and as nearly as possible, his temperature.

29. Who Are We?

A seasonal lack of demand for our product was responsible for a long period of hibernation for Alf, Bert, Charlie, Duggie, Ernie, Fred and George during the winter months. When they awoke they were in a state of only partial awareness. They knew *that* they were, but had forgotten who they were, what their occupations were and who their wives were.

But the subconscious worked wonders. Certain remarks were made, all in the third person, and any remark which anyone made in which his own name came was false, while anything else he said was true.

As Managing Director I see to it that individuals change their occupation from time to time so that my staff does not necessarily have now the same jobs as they had before. But the tasks that are to be done cannot be altered; they are Door-Knob-Polisher, Door-Opener, Door-Shutter, Bottle-Washer, Sweeper-Upper, Welfare Officer and Worker. The names of our employees' wives are Agnes, Beatrice, Clarissa, Diana, Ethel, Flossie and Gertie. No man has the same first letter in his name as his wife has. Below are 21 numbered remarks. Each man makes three, and the total of the numbers of each man's remarks is the same, except for Duggie and Fred who are each one out.

(1) Ernie is the Worker. (2) Fred is not the Door-Opener. (3) Gertie is married to the Welfare Officer. (4) Clarissa's husband is the Door-Knob-Polisher. (5) Bert is not the Door-Opener. (6) George is married to Diana. (7) Agnes is not married to Bert. (8) Duggie is the Worker. (9) Charlie's wife is Flossie. (10) Clarissa is married to Bert. (11) Duggie is married to Beatrice. (12) Alf is the Bottle-Washer. (13) Charlie is the Door-Opener. (14) George is the Door-Knob-Polisher. (15) George is not the Door-Shutter. (16) Charlie is the Door-Shutter. (17) Ernie is the Welfare Officer. (18) Ernie is married to Gertie. (19) The numbers of two of Bert's remarks are

perfect squares. (20) Charlie is the Bottle-Washer. (21) Fred is the Door-Opener.

Find out for each man his occupation, the name of his wife, and which remarks he makes.

30. The Sad Demise of Derrick Demmit

The Corpse of the not-very-well-known but very pure mathematician, Derrick Demmit, was found on the runway at Airley Airport. He had been struck on the head by a Log. He was last seen alive at 3:09 P.M., his body was discovered at 3:13 P.M.

Alf, Bert, Charlie, Duggie, Ernie, Fred and George had been associated with Demmit for a long time. They were all at the Airport that afternoon. It seems that one of them could no longer stand the insufferably condescending way in which Demmit had used them for his puzzles over the years. It may be taken as certain that one of them is guilty.

Airley is an isolated place, but it is connected by train, by road and by a helicopter service with the four towns, P, Q, R and S. The distances of these towns by road and air from Airley Airport (A.A.) are:

$$\begin{aligned} P &— A.A. & 40 \text{ miles} \\ Q &— A.A. & 60 \text{ "} \\ R &— A.A. & 80 \text{ "} \\ S &— A.A. & 90 \text{ "} \end{aligned}$$

Trains: Below is an extract from the train time table:

P.	2:30 P.M.	R.	1:50 P.M.
A.A. arr.	3:10 P.M.	A.A. arr.	3:14 P.M.
dep.	3:15 P.M.	dep.	3:20 P.M.
Q.	4:22 P.M.	S.	4:57 P.M.
Q.	1:59 P.M.	S.	1:26 P.M.
A.A. arr.	3:07 P.M.	A.A. arr.	3:04 P.M.
dep.	3:12 P.M.	dep.	3:08 P.M.
P.	3:54 P.M.	R.	4:36 P.M.

All trains run every half hour.

Helicopters: The times of the helicopters' departures are as follows:

P. to A.A.	2:34 P.M.	A.A. to P.	3:09 P.M.
Q. to A.A.	2:50 P.M.	A.A. to Q.	3:50 P.M.
R. to A.A.	2:10 P.M.	A.A. to R.	3:43 P.M.
S. to A.A.	2:58 P.M.	A.A. to S.	3:53 P.M.

The average speed of the helicopters is always exactly 80 m.p.h. and the service runs every hour (i.e. to find the times of the helicopters before or after those given subtract or add *one hour*).

Road: There are fine modern roads between P, Q, R, S and A.A. and an excellent car hire service. The average speed for anyone who uses this service is 50 m.p.h.

For various reasons the movements of our 7 suspects have been under observation for some time. Each of them was seen that afternoon in *two* of the places P, Q, R, S. It may be taken as certain that in travelling from one place to the other by whatever method, each of them passed through A.A.

They were seen as follows:
1. Alf was seen in P. at 2:32 P.M. and in S. at 4:54 P.M.
2. Bert was seen in Q. at 1:47 P.M. and in S. at 4:56 P.M. (Bert gets air sick and cannot travel by helicopter.)
3. Charlie was seen in R. at 2:09 P.M. and in P. at 4:02 P.M.
4. Duggie was seen in S. at 1:25 P.M. and in P. at 3:40 P.M.
5. Ernie was seen in S. at 1:56 P.M. and in R. at 4:41 P.M.
6. Fred was seen in Q. at 1:52 P.M. and in P. at 3:59 P.M.
7. George was seen in R. at 2:11 P.M. and in Q. at 4:37 P.M.

Find, for each of the seven suspects, the earliest time at which he could have arrived at A.A. and the latest time at which he could have left.

What can be said about which of them could have been guilty?

PART III

Hockey

31 — 43

31. 3 Teams

Three hockey teams A, B and C play against each other. The following table gives some information about the results.

	Played	Won	Lost	Tied	Goals for	Goals against
A	2	2				1
B	2			1	2	4
C	2				3	7

Complete the table and find the score in each game.

32. 5 Teams

In a competition between 5 hockey teams, A, B, C, D, E, each side is to play each of the others once.

The following table gives a certain amount of information as to what has happened after some of the games have been played.

	Played	Won	Lost	Tied	Goals for	Goals against
A	3	2	0		7	0
B	2	2	0		4	1
C	3	0	1		2	4
D	3	1	1		4	4
E	3	0			0	

Who has played whom? Find the result of each game and the score.

33. 6 Teams

6 hockey teams, A, B, C, D, E and F, are all to play each other once. After some of the games have been played a battered and inky piece of paper is discovered on which has obviously been written particulars of the results so far.

All that can be read is shown below:

	Played	Won	Lost	Tied	Goals for	Goals against	Points
A	2	1			4	2	
B	4				1	4	3
C			1			7	7
D	3				1	5	3
E						7	
F	5				2		7

2 points are awarded for a win; 1 point to each side for a tie; and no points for a match lost.

Find who played whom and the score in each match.

34. D Was Dumb

4 teams A, B, C and D are all to play one game of hockey against each other.

After some of the games had been played I managed to get particulars from the secretaries of A, B and C of the numbers of games which their teams had played, won, lost etc. But the secretary of D for some reason firmly refused to give any information at all.

The figures given were as follows:

	Played	Won	Lost	Tied	Goals for	Goals against
A	3	2	0	1	2	0
B	2	1	0	1	4	3
C	2	0	2	0	3	6
D						

Find the score in each game.

35. A Long Leg

I was interested to discover a piece of paper which appeared to give some details of the hockey matches that were being played in a competition between 3 local teams in which they were all to play each other once.

All that could be read was as follows:

	Played	Won	Lost	Tied	Goals for	Goals against	Points
A	1				6		
B	2	1			0	3	1
C				0	0	1	

I could see that apart from the missing figures there was something not quite right about this, so I took it along to a friend of mine who is secretary of one of the teams to see if he could help.

After looking at it briefly he laughed heartily and said, "Someone must have been pulling your leg; every single figure here is incorrect. But, no doubt," he added sarcastically, "you will be able to get what you want to know by the use of reason".

I discovered later that in fact from the information I had, which also included the fact that each team had played at least one game and that not more than 5 goals were scored in any game, it was possible to discover all details of the matches played. (2 points are given for a win, 1 point to each side for a tie and no points for a match lost.)

Find the score in each match.

36. Uncle Bungle Has His Moment

There must, I think, be some plot among his friends to leave lying around documents containing what appear to be tables of results of hockey matches for Uncle Bungle to discover.

He picked up another one the other day purporting to give some details of the games played between 3 local teams, A, B, and C.

It read as follows:

	Played	Won	Lost	Tied	Goals for	Goals against
A	2			1	4	3
B		0			1	2
C	1	0	0		2	2

It did not take Bungle long to discover on the basis of these figures, the score in each game. But it so happened that he *knew* the score in one game and the results which he got from this table did not agree with his knowledge. Bungle is getting rather suspicious now and he (quite rightly) came to the conclusion that *all* the figures in the table which he had found were incorrect. But nevertheless with the help of his knowledge of what had happened in one match (in which 2 goals were scored) he was able to discover all details of the correct table and the score in each game. He knew also that each side had played at least once and that not more than 4 goals were scored in any game.

(*i*) *What was the (incorrect) score in each game on the basis of the table given?* (*ii*) *What was the correct score in each game?*

37. Four Figures False

3 teams are to play each other once at hockey.

A table is discovered giving details of matches played, won, etc., as follows:

	Played	Won	Lost	Tied	Goals for	Goals against	Points
A	2	2	0	1	0	2	3
B	2	1	1	0	3	6	2
C	1	0	1	2	0	1	1

The secretary of one of the sides gave me the information that not all these figures were correct. In fact he said, and I am sure he was right, that 4 of them were wrong.

Find the correct table and the score in each match. (2 points are given for a game won, 1 point to each side for a tied game and no points for a lost game.)

38. Uncle Bungle Predicts

Uncle Bungle has become particularly interested lately in predictions. It was natural therefore that he should combine this with his life long interest in hockey.

4 local teams who were to play each other once had completed some of their games of which the detailed results can be deduced from the following (correct) table:

	Played	Won	Lost	Tied	Goals for	Goals against
A	1	1	0	0	3	1
B	2	1	0	1	3	1
C	2	0	2	0	1	5
D	1	0	0	1	1	1

Uncle Bungle was aware of these facts and he watched the teams frequently. On the basis of his observations and some tips about how to foretell the future which he had received from various friends, he drew up the following table giving the situation as he predicted it would be when all the matches had been played:

	Played	Won	Lost	Tied	Goals for	Goals against
A	3	2	0	1	6	1
B	3	1	1	1	3	4
C	3	0	3	0	1	6
D	3	1	0	2	2	1

Sadly, something had gone wrong. In fact Uncle Bungle was incorrect in his predictions of all the games that were still to be played. Not only was he incorrect about every result, but also about the number of goals scored by each side in each game. Furthermore the figures for goals for and goals against were incorrect for every team in his table.

In fact in the games about which Uncle Bungle was making predictions a total of 5 goals were scored.

Find the correct scores in all the games between A, B, C and D.

39. Some Ham Humor

It was a matter of some importance to me that I should know the details of the hockey games between 4 local teams, whom I will call A, B, C and D. They were all to play each other once but I did not even know whether all the matches had been played, though I knew that each team had played at least one match.

An old acquaintance of mine, who has a curious sense of humor and seems to think it amusing to deceive, had given me the following table of matches played, won, lost, drawn, etc., but I subsequently discovered that every single figure in this table was incorrect.

	Played	Won	Lost	Tied	Goals for	Goals against	Points
A	3	1	1	1	4	0	3
B	2	2	0	0	0	2	3
C	2	0	0	2	3	3	4
D	3	1	0	2	3	5	0

I found out from another — reliable — source that not more than 3 goals were scored in any game, and that only 1 goal was scored in a game in which B played. (2 points are given for a win, one point to each side for a tie and no points for a lost game.)

Find the correct table and the score in each game.

40. Uncle Bungle Baffled

Uncle Bungle picked up a tattered piece of paper containing what was obviously some particulars of the games played between 5 hockey teams.

It did not take him long to discover that there was something wrong about it but unfortunately he did not know whether all the figures were incorrect or just some of them. He was unable to get any further in his insatiable desire to discover precise details of the games. I was able to find out however that in fact only 1 figure was wrong.

The figures in the table which Uncle Bungle found were as follows:

	Played	Won	Lost	Tied	Goals for	Goals against
A	3	0		2	3	4
B	2		1	0	3	0
C	2		1		1	5
D				2		2
E	2		1		0	

Find which figure was incorrect, and the score in each game.

41. A Muddy Muddle

Uncle Bungle has been trying to find the exact results of a number of hockey games played between 6 teams (A, B, C, D, E, F) each of which is eventually to play each other once.

After some of the games had been played he managed to get hold of a piece of paper in which the numbers of games played, won, lost, tied etc., had been written out. But unfortunately my uncle then dropped this in the mud and all that could be read was as follows:

	Played	Won	Lost	Tied	Goals for	Goals against	Points
A	4			0		3	
B			3		6	5	3
C	3	2			2	5	
D	5	0		3	0	5	
E	3	2			7	4	6
F	5				4		6

I subsequently discovered that in fact one of these figures was incorrect. (2 points were given for a win, 1 point to each side for a tied game and no points for a game lost.)

Who played whom and what was the score in each game?

42. No Light of Truth

Generally Uncle Bungle only gets *some* things wrong, but occasionally, on his best — or worst — days no light of truth shines through.

He is a bit hard of hearing and does not see very well. These things were no doubt partly responsible for the fact that in a table which he produced giving the numbers of games played, won, lost, etc. by 5 local hockey teams, who were each to play each other once eventually, every single figure was incorrect.

I discovered from another source that each of these teams had played at least one game, that 4 games were tied and that not more than 3 goals were scored in any game.

Furthermore I managed to get from the secretary of B the information that in a game in which B played, one side scored 3 goals and that altogether B scored 2 goals more than C.

The table that Uncle Bungle produced was as follows:

	Played	Won	Lost	Tied	Goals for	Goals against	Points
A	4	0	0	4	5	1	2
B	3	3	0	0	3	4	1
C	4	4	0	0	2	1	1
D	3	0	3	0	2	3	3
E	4	0	3	1	4	0	2

(2 points are given for a win, 1 point to each side for a tied game and no points for a lost game.)

Find who played whom and the score in each game.

43. Zanies, Yodellers and X's

6 teams are all to play each other once at hockey. 2 of them belong to the XX Club, 2 to the Youthful Yodellers Club and the other 2 to the Zestful Zanies Club.

When some of the games had been played the secretaries of the 3 Clubs were asked to produce details of games played, won, etc. They all did so, with the following results:

XX Club

	Played	Won	Lost	Tied	Goals for	Goals against
A	4	1	1	2	7	6
B	3	0	2	1	0	7
C	5	1	3	1	1	2
D	4	1	0	3	6	2
E	3	2	0	1	4	1
F	5	0	5	0	1	7

Youthful Yodellers Club

	Played	Won	Lost	Tied	Goals for	Goals against
A	5	2	2	1	9	9
B	4	1	1	2	4	0
C	3	0	1	2	2	4
D	5	0	5	0	1	6
E	2	0	2	0	0	6
F	4	2	1	2	4	2

Zestful Zanies Club

	Played	Won	Lost	Tied	Goals for	Goals against
A	3	3	0	0	5	0
B	5	5	0	0	10	3
C	2	2	0	0	5	1
D	3	2	1	1	5	1
E	4	1	1	2	2	0
F	2	1	0	1	5	0

It will be seen that these 3 sets of results are different in every way.

In fact in the figures produced by the secretary of the XX Club, *all* details of the 2 teams belonging to the XX are correct, and all details of the other 4 teams are wrong; similarly in the figures produced by YY details of the 2 teams belonging to YY are right, and details of the other 4 teams are wrong; and in the figures produced by ZZ details of the 2 teams belonging to ZZ are correct, and details of the other 4 teams are incorrect.

Find who had played whom and the score in each game.

PART IV

Cross Number Puzzles

44 — 48

44. Straight as a Die

(There are no o's)

Across
1. A multiple of 11 across.
4. A prime number.
5. Odd.
6. Digits all different; their sum is 10.
9. Digits all different.
11. A prime number. The sum of the digits is more than 5 and less than 11.

Down
1. Odd. Each digit is greater than the preceding one.
2. Sum of digits is greater than 11.
3. The same two digits as in 4 across.
5. A perfect cube.
7. Even. Sum of digits is 19.
8. A perfect square.
10. A perfect cube.

45. Straight as a Cork-Screw

(*There are no o's*).

	1	2	
3			4
5			
		6	

Across
1. Sum of digits is 13.
3. Each digit differs by two from the preceding one in the same sense (i.e. all ascending or all descending).
5. Digits all even; their sum is 12.
6. A square.

Down
1. The sum of the digits is the same as the sum of the digits of 4 down.
2. The sum of the first two digits is the same as the sum of the last two digits of 1 down.
3. A cube.
4. The cube of an odd number.

46. Uncle Bungle's Cross Number

Uncle Bungle has been making up a cross number puzzle. But unfortunately, inevitably, something went wrong.

The puzzle, as he set it, was as follows:

1	2	3
4	5	6
7	8	9

Clues

Across

(1–2–3) The first digit is less than the second digit by the same amount as the second digit is less than the third.

(4–5–6) (Reversed). A square.

(7–8–9) The square of an odd number.

Down

(1–4–7) A factor of (2–5–8).

(2–5–8) Odd.

(6–9) A cube.

One clue is incorrect. Which one?

With which digit should each of the 9 squares be filled? (There are no 0's.)

47. A Three-Dimensional Cross Number

Top

1	2	3
4	5	6
7	8	9

Middle

10	11	12
13	14	15
16	17	18

Bottom

19	20	21
22	23	24
25	26	27

The 3 squares above are the 3 levels of a cube which is the basis of a 3-dimensional cross number puzzle. Each square is to be filled by a single digit. (*There are no o's.*)

Clues may be given at the same level — e.g. (11, 14, 17), or straight up or down — e.g. (3, 12, 21), or diagonally — e.g. (7, 17, 27) or (21, 14, 7).

In every case the clues will be of squares which are in a straight line.

Clues

Top

(1, 4, 7) Divisible by 3.
(1, 2, 3) Sum of digits = 17.
(3, 6, 9) A perfect square.
(7, 8, 9) Even. Sum of digits is greater by 2 than sum of digits of (3, 6, 9).

Middle

(13, 14) A perfect square.
(10, 13, 16) Sum of digits = 12.
(10, 11, 12) Each digit less than preceding one.

(12, 15, 18) An odd number.
(16, 17, 18) A multiple of 3.

Bottom

(21, 23, 25) An even number.
(22, 23, 24) Digits increase in size by the same positive amount.
(20, 23, 26) Each digit is greater than the one before. The sum of the digits is the same as the sum of the digits of (21, 24, 27).

Other clues

(7, 16, 25) A perfect cube.
(3, 12, 21) A multiple of (13, 14); sum of digits = 13.
(4, 13, 22) Digits ascend. 3rd digit divided by 2nd digit equals 2nd digit divided by 1st.
(5, 14, 23) Digits all different; their sum is *19*.
(9, 18, 27) Each digit is less than the preceding one.
(3, 15, 27) A perfect square.
(7, 17, 27) An odd number. The sum of the digits = 11.
(3, 14, 25) Each digit is greater than the preceding one.
(7, 13, 19) Difference between 1st and 2nd digits is equal to the difference between 2nd and 3rd digits in that order.

Find the numbers which should fill the 27 squares.

48. A Treble Cross Number

In this cross number puzzle there are three different sets of solutions which fit the clues given. You are asked to find them all. In some cases the same square may be filled by the same digit in different solutions, *but in no case is the complete answer to a clue the same in different solutions.* (*There are no o's.*)

Across

1. The sum of the digits is the same as the sum of the last 3 digits of 6 down.
3. A multiple of 16.
5. The sum of the digits is half the sum of the digits of 7 across.
7. The first digit is greater than the second digit by the same amount (which may be 0) as the second digit is greater than the third.
9. Digits all different — none greater than 5.
11. A perfect square.
12. Half of 3 across.

Down

2. (Reversed.) A multiple of 9.
3. Sum of digits equals 16.

4. Number formed by first two digits is twice the number formed by the last two.
6. Successive digits increase either by the same amount or in the same proportion (i.e. either in an Arithmetic or a Geometric progression).
8. A perfect cube.
10. The magic number.

Find the 3 solutions and hence the 3 (different) magic numbers (see 10 down).

PART V

Assorted — Not So Simple

49 — 62

49. Top Hats and Tails

The following notice goes up at Help-the-Boys Hall:

Clothing Regulations
1. Only if a boy is over 16 may he wear a tail coat.
2. No boy who is not over 15 may wear a top hat.
3. The wearing of either a top hat or a tail coat or both is a necessary condition for watching baseball on Saturday afternoon.
4. If a boy is either carrying an umbrella or over 16 or both he must not wear a sweater.
5. Boys must either not watch baseball or wear a sweater or both.

What can you say about the age and apparel of those watching baseball on Saturday afternoons?

50. The Older the Better

A, B and C are talking about their ages, which are all different and all between 31 and 43 inclusive.

They each make 2 remarks, as follows:
A: 1. B is 3 years older than I am.
    2. C is 4 years younger than I am.
B: 1. A's age is a multiple of 6.
    2. C is older than A.
C: 1. I am one year older than A.
    2. A's age is a multiple of 7.

An analysis of these remarks showed the interesting fact that the older a man is the more truthful he is. The oldest of the three made two true remarks, the youngest made none, and the other made one true and one false remark.

Find out as much as you can about their ages.

51. Orders About Orders

A, B, C, D, E are five quantities which have a numerical value. You are told that:
1. A is equal to C if and only if E is not equal to B.
2. Only if C is as much less than B as B is less than A, is A greater than D.
3. C is less than A and greater than D.

What is the order of magnitude of A, B, C, D, E?

52. Who's Sitting Where?

Mr. and Mrs. Binks, Mr. and Mrs. Bloggs and Mr. and Mrs. Bunn go to the theatre together; they sit side by side in one row with ladies and gentlemen in alternate seats. No man sits next to his wife. The men's names are John, Rupert and Ethelred; their jobs are dentist, accountant and principal; the names of their wives are Jane, Elizabeth, Dawn. (These names and jobs are in no particular order.)

The dentist occupies one of the middle two seats and he is sitting nearer to Dawn than he is to Mr. Bloggs or to John.

Mr. Binks is at one end of the row and he has the principal's wife on his right.

Rupert is sitting between Jane, on his left, and Elizabeth.

Find out where they are all sitting, their Christian names and the men's jobs.

53. In-Laws

Five people, A, B, C, D, E are related to each other. Four of them make one true statement each as follows:
 1. B is my father's brother.
 2. E is my mother-in-law.
 3. C is my son-in-law's brother.
 4. A is my brother's wife.

Each person who is mentioned is one of the five (e.g. when someone says 'B is my father's brother' you can be sure that 'my father' as well as 'my father's brother' is one of A, B, C, D, E).

Find out who made each of the four statements and how the five people are related.

54. Only Two True

A, B, C, D, E, F, G are arranged in a certain order (no ties) as the result of a competition. Statements are made as follows:
 1. E was 2nd or 3rd.
 2. C was not 4 places higher than E.
 3. A was lower than B.
 4. B was not 2 places lower than G.
 5. B was not first.
 6. D was not 3 places lower than E.
 7. A was not 6 places higher than F.

Only two of these statements are true. Which? Find the order.

55. Some Buffers Bite

In a certain community the membership of the Buffer Club is composed of all those over 70 who are not toothless, of all toothless people who are not members of the Nashum Club and of all members of the Nashum Club who are not over 70.

Can you tell whether a man is a member of the Nashum Club if you are told he is:
  (i) over 70, toothless, not a member of the Buffer Club?
  (ii) under 70, possessed of teeth, not a member of the Buffer Club?
  (iii) under 70, toothless, a member of the Buffer Club?

56. The Island of Perfection

On the Island of Perfection after years of scientific experiment and research they have at last won the battle against the forces of gravity. They have discovered a material which is not merely weightless but has a negative weight. Naturally a first thought of the inventors is how nice it will be to be able to lessen the weights and therefore the cost of their letters and parcels. But even on the Island of Perfection one cannot expect to get less than nothing for nothing and the stamps which are made of the special new material, upsy, are expensive.

The unit of currency on the island is the Perfect Dollar (P.D.) and the unit of weight is the *Bounce* (B).

In the early days of the invention 3 different kinds of stamps were available — the old-fashioned pre-upsy, *black* stamps of negligible positive weight; the upsy *green* stamps which weigh *minus* 1 B for each 2 P.D. stamp; and the super-upsy *red* stamps which weigh *minus* 2 B for each 2 P.D. stamp. Every stamp is a 2 P.D. of one of the three colors.

The postal regulations for stamps of different colors are set out as follows:

Weight	Black stamps	Green stamps	Red stamps
not exceeding 3 B	2 P.D.	4 P.D.	6 P.D.
over 3 B, but not exceeding 8 B	4 P.D.	8 P.D.	10 P.D.
each additional B thereafter	2 P.D.	4 P.D.	6 P.D.

On every letter or parcel sent there may not be stamps of more than one color.

What is the cheapest way of sending letters or parcels of the following weights:

(i) 5 B; (ii) $9\frac{1}{2}$ B; (iii) $10\frac{1}{2}$ B; (iv) $13\frac{1}{2}$ B; (v) $15\frac{1}{2}$ B?

(Indicate in each case which stamps are being used. If there are two or more equally cheap alternatives give them all.)

57. Round the Table

Mr. and Mrs. Smith, Mr. and Mrs. Brown, and Mr. and Mrs. Green, are seated equally spaced round a circular table. No man is sitting next to his wife, but each lady has a man on each side of her.

The names of the men and their wives, not necessarily respectively, are Tom, Dick and Harry, and Nancy, Joan and Mary. The occupations of the men, again not necessarily respectively, are architect, politician and machine operator.

Dick and Mr. Smith often play bridge with the architect's wife and Mrs. Green.

The machine operator, who is an only child, has Mary on his right.

The politician is sitting nearer to Nancy than he is to Mrs. Brown.

Harry is the architect's brother-in-law, and he has his only sister sitting on his left. The architect is sisterless.

Find their names, their occupations, and the order in which they are sitting round the table.

58. The Family Tree of Truth

John, James, Nancy, Lucy and Pamela made statements as follows:

JOHN: Nancy is my wife; James is my son; Pamela is my aunt.

JAMES: Lucy is my sister; Pamela is my mother; Pamela is John's sister.

NANCY: I have no brothers or sisters; John is my son; John has a son.

LUCY: I have no children; Nancy is my sister; John is my brother.

PAMELA: John is my nephew; Lucy is my niece; Nancy is my daughter.

Assume that
1. anyone who has one or more brothers or sisters and one or more children always tells the truth;
2. anyone who has either one or more brothers or sisters or one or more children makes statements which are alternately true and false;
3. anyone who has no brothers or sisters and no children never tells the truth.

Find which of the above statements are true and what relations these five people are to each other.

59. The Dowells Do Better

The male members of the Dowell family are in the habit of having a dinner party from time to time. At the last of these festive occasions, held to celebrate the fact that they had been doing even better than usual, there were 5 of them present, (A, B, C, D and E), representing three generations, and they were sitting equally spaced round a circular table.

The study of circular dinners is a hobby of mine, so I persuaded them to give me some more information about their relationships with each other and their relative positions on this particular occasion.

They spoke as follows:
A: 1. I was sitting next to E.
B: 1. C is my nephew.
   2. I was sitting next to my father.
C: 1. D is my father.
D: 1. E was sitting next to me.
E: 1. I was sitting nearer to C than to A.
   2. C is sitting next to his brother.

I discovered afterwards that any members of the youngest of the 3 generations, perhaps as a result of being unaccustomed to the amount of hard liquor that it was customary to dispense on these occasions, were incorrect in any statements they might make about the relative *positions* of those dining.

Find how these 5 are related to each other and how they were seated round the table.

60. Logic about the Logic Prize

Smith, Brown, Jones and Robinson are told that they have each won one of the four prizes for Mathematics, English, French and Logic, but none of them know which. They are speculating about it. Smith says he thinks Robinson has won the Logic prize. Brown thinks that Jones has won the English prize. Jones feels confident that Smith has not won the Mathematics prize, and Robinson is of the opinion that Brown has won the French prize. It turns out that the winners of the Mathematics and Logic prizes were correct in their speculations, but the other two were wrong.

Who won which prize?

61. Some Eccentric Sportsmen

Two of the necessary qualifications for membership of the Eccentric Sportsmens' Club are that one should have played Polo in Patagonia and Croquet in Czechoslovakia. In order to get a permit to play Croquet in Czechoslovakia one must be a founder member of the Hoop Club. One cannot be a founder member of the Hoop Club unless one has played Polo in Patagonia. Everybody must either be a member of the Eccentric Sportsmens' Club or the Oddfellows' Association or both, but one cannot both be a member of the Oddfellows' Association and have played Croquet in Czechoslovakia.

What can you say about (i) Smith, who is not a founder member of the Hoop Club; (ii) Jones, who has played Croquet in Czechoslovakia; (iii) Robinson, who has not played Polo in Patagonia?

62. Happy Family Unions

In the Brown, Green, Black and White Houses live, but not respectively, the Brown, Green, Black and White families. No family lives in the house that bears its name. There is a son and daughter in each family and each son is engaged to one of the daughters, but no man is engaged to a lady who bears the same name as his family's house, and no lady is engaged to a man who bears the same name as her family's house. Mr. Black's fiancée does not live at the Brown House. Mr. Brown, together with the brother of his fiancée, and Mr. Green, were entertained at tea at the White House.

Where does everybody live? Who is engaged to whom?

PART VI

Missing Digits

63 — 72

63. Multiplication and Division

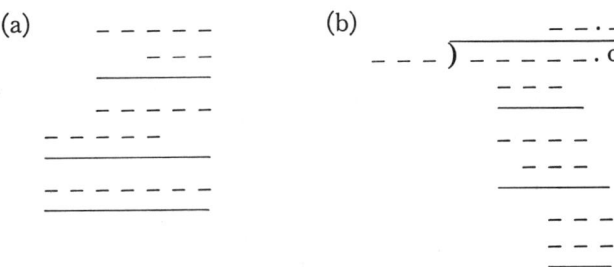

(a) and (b) represent the same two numbers in one case multiplied together, in the other case divided.

Given that there are no 3's, find all the missing digits.

64. 9 Digits Divided by 2 Digits

A long division sum:

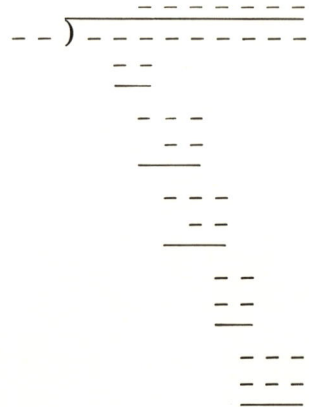

In the complete solution there are four 5's.
Find the missing digits.

65. 7 Digits Divided by 2 Digits

A long division sum:

```
           _ _ _ _ _._
    _ _ ) _ _ _ _ _ _ _
          _ _
          ___
          _ _ _
           _ _
           ___
            _ _ _
            _ _ _
            _____
              _ _
              _ _
              ___
               _ _ _
               _ _ _
               _____
                 ___
```

Find the missing digits.

66. 5 Digits Divided by 3 Digits

A long division sum:

```
            _ _._ _
       _____
 _ _ _ ) _ _ _ _ _
         _ _ _
         _____
           _ _ _
           _ _ _
           _____
             _ _ _
             _ _ _
             =====
```

Find the missing digits.

67. An Ascending Quotient

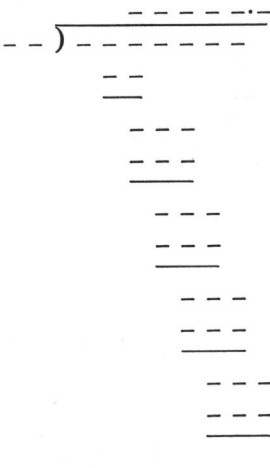

In this division the figures in the quotient ascend from left to right, apart from any o's there may be. There are two possible solutions.

Find them both.

68. An Unconvincing Division

Uncle Bungle had been looking at some of those long division sums with all the figures missing and it had occurred to him that greater verisimilitude would be given to the project if, instead of a blank, there was an incorrect figure.

In the following, obviously incorrect, long division the pattern is correct, but every single figure is wrong.

```
              1 9 3
        ┌──────────
   6 6 ) 9 1 2 6
        6 8
        ─────
        1 3 2
        1 1 1
        ─────
            2 1 5
            2 1 0
            ─────
```

Find the correct figures. (The correct division sum of course comes out exactly.)

69. 4 Digits Divided by 2 Digits (all incorrect)

In the following, obviously incorrect, division sum the pattern is correct, but every single figure is wrong:

$$
\begin{array}{r}
1\,7\,9 \\
79\,)\,9\,3\,9\,8 \\
6\,6 \\ \hline
1\,8\,9 \\
1\,5\,6 \\ \hline
1\,3\,5 \\
1\,3\,0 \\ \hline
\end{array}
$$

Find the correct figures. (*The correct division sum comes out exactly.*)

Solution: $8046 \div 27 = 298$

70. 6 Digits Divided by 2 Digits (all incorrect)

In the following, obviously incorrect, division sum the pattern is correct, but every single figure is wrong:

```
              9 2 1 3
      4 3 ) 2 3 8 9 9 5
            1 0 2
            ─────
                6 9
                6 4
                ───
                  5 9
                  2 8
                  ───
                    3 1 5
                    2 1 0
                    ═════
```

Find the correct figures. (The correct division comes out exactly.)

71. A Mistake in the Mistakes

Uncle Bungle has been making up another long division sum with all the figures wrong, but unfortunately he made a mistake and *one* of the figures is correct.

The puzzle, as he set it, looked like this:

```
              2 3 4 3
        1 8 ) 8 1 0 1 2
              6 9
              ___
              2 2 0
                8 0
                ___
                2 0 9
                  8 2
                  ___
                    7 2
                    7 2
                    ===
```

The pattern is that of a long division sum with all the figures incorrect except one.

Which one? Find all the correct figures.

93

72. Addition and Subtraction

Diagrams (i) and (ii) represent the same two sets of figures added together and subtracted.

(i) _ _ _ _ _
    6 _ _ 9
_ _ _ _ _ _

(ii) _ 4 _ _ _
    _ _ _ _
9 _ _ _ 3

In the upper number each digit is less than the preceding one, and in the second number each digit is greater than the preceding one. There are no 0's in either of the two numbers.

The figures given are all wrong.

Find the two numbers.

PART VII

The Island of Imperfection

73 — 92

73. Unable, Unhappy and Unwanted

There are three tribes on the Island of Imperfection — the Pukkas, who always tell the truth; the Wotta-Woppas, who never tell the truth; and the Shilli-Shallas, who make statements which are alternately true and false (or false and true).

The heads of the three tribes, whose names, in no particular order, are Unable, Unhappy and Unwanted, are giving a demonstration of tribal truthfulness.

UNABLE says: 1. Unwanted is a Shilli-Shalla.
2. I am a Pukka.
UNHAPPY says: Unable is a Wotta-Woppa.
UNWANTED says: Unhappy is a Wotta-Woppa.

Find to which tribe each of the three belong.

74. The Silent C

On the Island of Imperfection there are three different tribes — the Pukkas, who always tell the truth; the Wotta-Woppas, who never tell the truth; and the Shilli-Shallas, who make statements which are alternately true and false, or false and true.

A discussion group, with one member from each tribe (A, B and C) is practicing tribal characteristics.

A says: 1. I am a Shilli-Shalla.
2. B has said that he is a Pukka.
3. C has said that he is a Wotta-Woppa.
B says: 1. I am a Shilli-Shalla.
C is silent.

To which tribe do A, B and C belong?

75. The Charming Chieftainesses

There was a time when the leaders of the three tribes on the Island of Imperfection were all ladies. Their names were Attractive, Delectable and Fascinating and they had been having between themselves a competition in charm.

The tribes, of course, are the Pukkas, who always tell the truth; the Wotta-Woppas, who never tell the truth; and the Shilli-Shallas, who make statements which are alternately true and false (or false and true).

The three ladies made statements as follows:

ATTRACTIVE: 1. Delectable is a Shilli-Shalla.
2. Delectable is more charming than Fascinating.

DELECTABLE: 1. Fascinating is a Wotta-Woppa.
2. Attractive is a Pukka.

FASCINATING: 1. The most truthful of us is the least charming.
2. Attractive is certainly not the most charming of the three of us.

Find to which tribe each lady belongs, and the order in which they came in the charm competition.

76. Ringing Is Not All Roses

The inhabitants of the Island of Imperfection have recently been introduced to the delights of the telephone. These delights however are made somewhat complicated and uncertain by communication difficulties. There are 3 tribes on the island — the Pukkas, who always tell the truth, the Wotta-Woppas, who never tell the truth, and the Shilli-Shallas, who make statements which are alternately true and false, or false and true.

These tribal rules apply to statements about and the dialing of telephone numbers, all of which have 3 figures. Thus if a Wotta-Woppa makes a statement about a number all 3 digits will be incorrect. If a Shilli-Shalla makes a statement the digits will be alternately true and false, or false and true, and this alternation will be carried over from one statement to the next. And of course the figures in a Pukka's statement will all be correct.

3 inhabitants of the island, one from each tribe, make statements as follows:

A: 1. My number is 468.
    2. C's number is 403.
B: 1. My number is 942.
    2. A's number is 587.
C: My number is 304.

It so happened that any digit that was incorrect in these particular statements was 1 out either way, i.e., 1 more or 1 less than the correct digit.

Find to which tribe each man belongs. Give as much information as you can about their telephone numbers.

77. Logic Lane

On the Island of Imperfection there is a special road, Logic Lane, on which the houses are usually reserved for the more mathematical inhabitants.

Add, Divide and Even live in three different houses on this road (which has houses numbered from 1–50). One of them is a member of the Pukka Tribe, who always tell the truth; another is a member of the Wotta-Woppa Tribe, who never tell the truth; and the third is a member of the Shilli-Shalla Tribe, who make statements which are alternately true and false, or false and true.

They make statements as follows:

ADD:
1. The number of my house is greater than that of Divide's.
2. My number is divisible by 4.
3. Even's number differs by 13 from that of one of the others.

DIVIDE:
1. Add's number is divisible by 12.
2. My number is 37.
3. Even's number is even.

EVEN:
1. No one's number is divisible by 10.
2. My number is 30.
3. Add's number is divisible by 3.

Find to which tribe each of them belongs, and the number of each of their houses.

78. The Top Teams on the Imperfect Island

The three tribes on the Island of Imperfection have each produced their best possible soccer teams to play each other once.

After the games are over the 3 captains (A, B, C in no particular order) make some remarks about the games and about each other. These remarks are of course in accordance with the tribal rules — the captain of the Pukkas makes statements which are all true, the captain of the Wotta-Woppas makes statements which are all false, and the captain of the Shilli-Shallas makes statements which are alternately true and false (or false and true).

A says: 1. C's team scored 4 goals.
          2. C is a Wotta-Woppa.
          3. B's team only scored 1 goal.

B says: 1. A is a Pukka.
          2. We won both our games.
          3. Our team tied with C's team.

C says: 1. The Pukkas beat the Shilli-Shallas.
          2. A's team scored 3 goals against our team.
          3. B says that A is a Pukka.
          4. One game was tied.

Find to which tribe A, B and C belong, and the scores in the soccer games.

79. Some Tribal Wedlocks

Arthur, Bartholomew and Clarence are three inhabitants of the Island of Imperfection who are married, but not necessarily respectively, to Dulcy, Ermintrude and Fanny. All the inhabitants of the island are members of one of the three tribes — the Pukkas, who always tell the truth, the Wotta-Woppas, who never tell the truth, and the Shilli-Shallas, who make statements which are alternately true and false (or false and true).

Each of the three men belongs to a different tribe, as does each of the three women. It is a very strict law of the island that members of the same tribe shall not mate.

The three men make statements as follows:

ARTHUR: 1. Fanny is my wife.
2. Clarence is married to a Shilli-Shalla.
BARTHOLOMEW: 1. Clarence is not married to Fanny.
2. Ermintrude is not a Pukka.
CLARENCE: 1. Arthur is a Pukka.
2. Bartholomew is married to Fanny.
3. Dulcy is my wife.

Find to which tribe each of the six people belong and who is married to whom.

80. The Immortal Wolla

Things have been moving fast on the Island of Imperfection and the leaders have been getting a taste of the methods of the great world outside. In particular there has been a trend to adopt modern economic methods and each of the three tribes has appointed a Secretary of the Treasury. (The three tribes are the Pukkas, who always tell the truth; the Wotta-Woppas, who never tell the truth, and the Shilli-Shallas, who make statements which are alternately true and false — or false and true.) It has seemed to these Secretaries desirable to introduce a money system and the currencies, in no particular order, are to be Blanks, Wollas, and Mounds. It was a matter of some difficulty to decide on the rates of exchange between these currencies, but eventually agreement was reached. (In no case were the values exactly the same.)

The 3 Secretaries (whom we shall call A, B and C in no particular order) made statements to the press, in accordance, of course, with their tribal rules. As follows:

A: 1. 2 Wollas are worth 5 Mounds.
    2. Our currency is Blanks.
    3. The Wotta-Woppas' currency is the Wolla.

B: 1. A is a Pukka.
    2. 3 Mounds are worth 4 Blanks.
    3. The Shilli-Shallas' currency is worth more than the Wotta-Woppas'.

C: 1. B's currency is worth less than A's.
    2. 1 Blank is worth 3 Wollas.
    3. Our currency is the Wolla.

Find to which tribes A, B and C belong, the name of each tribe's currency and the rates of exchange among the currencies.

81. Imperfect Wives and Daughters

Arthur, Brian and Clarence live on the Island of Imperfection and each is therefore a member of one of the three tribes — the Pukkas, who always tell the truth, the Wotta-Woppas, who never tell the truth, and the Shilli-Shallas, who make statements which are alternately true and false, or false and true. Each one of the three belongs to a different tribe. They each have a wife and a daughter on the island, who all belong to one of the tribes. At the time with which this story deals there were no rules about tribal distribution within a family, but it so happens that in none of these 3 families do husband, wife and daughter all belong to the same tribe.

Arthur, Brian and Clarence make statements as follows:

ARTHUR:
1. Brian's daughter is a Pukka.
2. Clarence and his daughter belong to different tribes.
3. My wife is a Shilli-Shalla.

BRIAN:
1. Arthur is a Pukka.
2. My wife belongs to the same tribe as I do.
3. Arthur's wife and daughter belong to different tribes.

CLARENCE:
1. Arthur says that his daughter is a Shilli-Shalla.
2. Brian is a Shilli-Shalla.
3. I am not married to a Wotta-Woppa.

Find to which tribe each man, his wife and his daughter belong.

82. Competitions in Imperfection

A, B and C are three inhabitants of the Island of Imperfection. One of them is a member of the Pukka Tribe, who always tell the truth; another is a member of the Wotta-Woppa Tribe, who never tell the truth; and the third is a member of the Shilli-Shalla Tribe, who make statements which are alternately true and false (or false and true).

They have been competing in various kinds of imperfection — to see who is the most stupid, the plainest, and the most socially unacceptable, and they are placed in an order of merit (no ties).

After these competitions are over they each make three statements, which are, of course, in accordance with their tribal rules.

A: 1. B is higher in the stupidity test than in the socially unacceptable test.
    2. C is lower in the socially unacceptable test than in the plainness test.
    3. I occupy the same place in the socially unacceptable test as I do in the plainness test.

B: 1. I am not a Shilli-Shalla.
    2. I am less socially acceptable than C.
    3. C is a Pukka.

C: 1. A is the most socially acceptable of the three of us.
    2. A is a Wotta-Woppa.
    3. I am more stupid than A.

Find to which tribes A, B and C belong and the order of merit in the three tests.

83. Uncle Bungle on the Island

Uncle Bungle has at last fulfilled a long-standing ambition and has found his way to the Island of Imperfection. He is aware of the fact that everyone on the island belongs to one of 3 tribes — the Pukkas who always tell the truth, the Wotta-Woppas who never tell the truth, and the Shilli-Shallas who make statements which are alternately true and false, or false and true. But unfortunately Uncle Bungle is color-blind and is unable to distinguish correctly between the different tribal trousers; in fact he gets them all wrong.

It was I think, just to exercise his logic and to display his knowledge of their customs that my uncle asked 3 natives (A, B and C) about the tribes of each others' wives.

They answered as follows:
A: C's wife is a Wotta-Woppa.
B: A's wife is a Pukka.
C: B's wife is a Wotta-Woppa.

After a pause for reflection Uncle Bungle said triumphantly:

"B's wife is a Pukka, and I know to which tribes the other wives belong."

He was in fact wrong about the tribe of all 3 wives, but his eyes were good enough for him to realize, correctly, that the 3 men all belonged to different tribes. His assumption that the wives belonged to different tribes was also correct, but this, I think, was sheer good fortune.

My uncle assumed for some reason that the Shilli-Shalla was due to make a false statement, but in this he was incorrect.

At the time of his visit the only law on the island about the inhabitants marrying or not marrying members of the same tribe was that in order that the light of truth should not be entirely extinguished a union between two Wotta-Woppas was not allowed. My uncle was aware of this law.

1. *To which tribes did Uncle Bungle think that A, B, C and their wives belonged?*
2. *To which tribes did they in fact belong?*

84. The Tribal Trousers

The Island of Imperfection is now so 'with it' that everyone wears trousers, with the result that it is sometimes difficult to distinguish the sexes. And as we shall be referring to the six people (3 married couples) with whom this story is concerned by the letters A, B, C, D, E, F their names will not help either.

They are all, of course, members of one of the tribes on the island — the Pukkas, who always tell the truth, the Wotta-Woppas who never tell the truth, or the Shilli-Shallas who make statements which are alternately true and false (or false and true). And their trousers bear their tribal colors. One of the laws of the Island prevents husband and wife from belonging to the same tribe.

The ages of these six people are all different; none of them is older than 46 or younger than 29.

They make statements as follows:

A: 1. My age is a multiple of 7.
   2. I am not married to E.
   3. F is two years younger than D.

B: 1. B is older than C.
   2. D is not a Wotta-Woppa.
   3. E is a Pukka.

C: 1. E is 10 years older than C.
   2. B is 6 years older than F.
   3. I am married to the oldest person.

D: 1. C is not a Pukka.
   2. F is a Pukka.
   3. B is older than C.

E: 1. F is not a Pukka.
   2. C is 5 years older than A.
   3. D is a Pukka.

F: 1. B is a Shilli-Shalla.
   2. E is a Wotta-Woppa.
   3. I am married to D.

Find to which tribe each person belongs, who is married to whom, and their ages.

85. Some Imperfect Wages and Taxes

Since the introduction of economic theory to the Island of Imperfection much of their talk and thoughts is about money. Members of the Shilli-Shalla tribe who make statements which are alternately true and false, or false and true, have been particularly interested and our story is about the remarks which three of them — Rockerfeller, Scrooge and Tight — made in conversation with each other. They have learned the lesson, so important in an advanced economy, of not giving away anything, and the remarks which they make are anonymous under the headings A, B and C.

These remarks are as follows:

A: 1. My income before tax is neither w 372 nor w 376 (w stands for Wollas, the Shilli-Shallas' units of currency).
    2. B is Scrooge.
    3. I pay 10% of my income in tax.

B: 1. C is Rockerfeller.
    2. C's income before tax is either w 440 or w 450.
    3. I am not Tight.

C: 1. B's income after tax is w 321.
    2. I pay one third of my income in tax.
    3. A is Tight.

These remarks are of course in accordance with the tribal rules; it is interesting to notice that two of them start with a true remark and one with a false.

Their wages and taxes are all an exact number of w's. No one gets more than w 450.

In accordance with the general imperfection of the island the higher a man's salary the lower the percentage of it which is deducted in tax. The percentage tax deductions of these three are all different, one is 10%, another is 25% and the third is $33\frac{1}{3}$%.

Find who A, B and C are, the wages which they receive and the amount which they pay in taxation.

86. The Years Pass

Even on the Island of Imperfection — or perhaps especially on the Island of Imperfection — no one gets any younger. Four inhabitants of the island were having a conversation, as they so often did, about their tribes and they were also making various statements about their ages. (I happen to know that all of them are over 12.)

There are 3 tribes on the Island — the Pukkas who always tell the truth, the Wotta-Woppas who never tell the truth, and the Shilli-Shallas who make statements which are alternately true and false or false and true. (But any Shilli-Shalla there may be among these four starts with a true statement.) There is at least one representative of each tribe among these four.

They speak as follows:

A: 1. B and D belong to different tribes.
    2. C is 11 years older than D.
    3. B is 34.
B: 1. The oldest of us is a Pukka.
    2. D's age is a multiple of 9.
    3. C is a Pukka.

C: 1. D is a Pukka.
    2. I am 5 years younger than B.
    3. A is a Wotta-Woppa.
D: 1. C's age is a prime number.
    2. B is 29.
    3. A is 1 year older than C.

Find the age of each man and the tribe to which he belongs.

87. A Mixed Marriage

Our story deals with a marriage on the Island of Imperfection which took place many years ago now. It was between a member of the Pukka Tribe (who always tell the truth) and a member of the Wotta-Woppa Tribe (who never tell the truth), and the union was blessed with a son who naturally grew up with the characteristics of the Shilli-Shalla Tribe (who make statements which are alternately true and false, or false and true).

So happy was this union that each spouse yielded over the years to the characteristics of the other, and at the time with which this story deals the Pukka was in the habit of making one false statement for every three true ones, and the Wotta-Woppa was in the habit of making one true statement for every three false ones.

These two parents and their son each have tribal numbers, which are all different, and their names, in no particular order, are Cecil, Evelyn and Sidney (these names are all used on the island for members of either sex).

Each of the three makes 4 remarks — but they are made anonymously so that we have to deduce by whom each set was made. (It may be taken for granted that the ex-Pukka makes 1 false and 3 true remarks, and the ex-Wotta-Woppa makes 1 true and 3 false.)

Their remarks are:

A: 1. Cecil's number is the greatest of the three.
    2. I am an ex-Pukka.
    3. B is my wife.
    4. My number is greater than B's by 22.

B: 1. A is my son.
    2. My name is Cecil.
    3. C's number is either 54 or 78 or 81.
    4. C is an ex-Wotta-Woppa.

C: 1. Evelyn's number is greater than Sidney's by 10.
    2. A is my father.
    3. A's number is either 66 or 68 or 103.
    4. B is an ex-Pukka.

Find out which of A, B and C is mother, father and son; the names of each of them and their tribal numbers.

88. Uncle Bungle Fails Again

Uncle Bungle has been on the Island of Imperfection for some time now, and he is very keen to join one of the three tribes — the Pukkas, who always tell the truth, the Wotta-Woppas who never tell the truth, or the Shilli-Shallas, who make statements which are alternately true and false or false and true. He has his opinion as to which tribe he would prefer, but above all he wants to be accepted. After much negotiation he manages to get the authorities to agree to his having a tribal test. With due pomp and ceremony he and a leading member of each tribe are to make 3 remarks each.

Unfortunately Uncle Bungle failed in his test. His remarks did not conform with the specifications required for any one of the tribes. And it was a sad uncle who eventually left the tribal arena, perhaps forever.

The precise manner of failing was of course recorded and it may be of interest to set out the details for posterity.

The remarks made by the four of them were as follows:

A: 1. B is married to Gertie.
    2. B is a Wotta-Woppa.
    3. C is married to a Wotta-Woppa.
B: 1. Gertie is not a Wotta-Woppa.

  2. C is Uncle Bungle.
  3. D is married to Florinda.
C: 1. Florinda is not a Wotta-Woppa.
  2. B's wife is a Wotta-Woppa.
  3. I am a Shilli-Shalla.
D: 1. C is a Shilli-Shalla.
  2. Ellie and Gertie belong to the same tribe.
  3. A is a Pukka.

Uncle Bungle has no wife but the other three are married to Ellie, Florinda and Gertie who are all, of course, members of one of the tribes, not necessarily all different. Not one of the ladies belongs to the same tribe as her husband.

Find which of A, B, C and D is a Pukka, which a Shilli-Shalla, which a Wotta-Woppa and which is Uncle Bungle.

Find also the tribes of the 3 ladies and to whom they are married.

89. World Cup for the Wotta-Woppas?

All the tribes on the Island of Imperfection have become keen soccer players, but none of them has become so enthusiastic as the Wotta-Woppas. Four teams (A, B, C and D) have been competing for the honor of representing the tribe in the World Cup (they are each to play against each other once), and I was fortunate enough to come across a document giving some details of how the competition was going after some of the matches had been played.

The document read as follows:

	Played	Won	Lost	Tied	Goals for	Goals against
A					6	2
B	2	2	0	2	4	1
C	3	1	2	0	0	1
D	1	1	2	0	0	2

The fact that some of the figures were missing did not help in my attempts to discover the results of all the matches played; nor did the fact that in accordance with the rules of the Wotta-Woppas *all* the figures given were incorrect. I was lucky enough to discover however that whether by accident or design each figure was exactly 1 out, i.e., 1 more or 1 less than the correct figure.

Find who played whom and the score in each game.

90. Some Assorted Tribal Soccer

Enthusiasm about football on the Island of Imperfection has been growing fast. At the time of writing there is one league of which the members are competing keenly against each other with some matches still to be played. (They are all to play against each other once.)

In this league there are 4 teams, one representing the Pukka Tribe, who always tell the truth; one representing the Wotta-Woppa Tribe, who never tell the truth; one representing the Shilli-Shalla Tribe, who make statements which are alternately true and false; and one assorted team with members of all 3 tribes.

I managed to get some information from the secretary of each team about the numbers of matches played, won, lost, tied etc., but naturally the figures given me were in accordance with tribal characteristics — that is to say the figures given me by the Pukka Tribe will all be true, those given me by the Wotta-Woppas will all be false, those given me by the Shilli-Shallas will be alternately true and false (or false and true), and those given me by the secretary of the assorted team will be some true and some false in no particular order.

Looking back at the events later it was interesting to see that when the figures were wrong they were in each case either one more or one less than the correct figures.

The figures given by the four secretaries were as follows. (2 points are given for a win, and 1 for a tie.)

	Played	Won	Lost	Tied	Goals for	Goals against	Points
A	3	0	0	3	5	4	3
B	3	0	2	1	0	5	2
C	2	1	1	0	4	4	2
D	2	0	2	0	3	2	2

Find the correct figures, and the score in each game.

91. Our Factory on the Island

Alf, Bert, Charlie, Duggie and Ernie have recently had a well-earned raise in salary and they decided to spend their vacation together, with their wives, on a cruise through some of the less known waters of the world. Their wives' names are, not necessarily respectively (and — sadly — not necessarily as they used to be), Priscilla, Queenie, Rachel, Sarah and Tess.

Unfortunately they ran into bad weather and were shipwrecked on the Island of Imperfection. There are now five tribes on this island, the Pukkas, who always tell the truth; the Wotta-Woppas, who never tell the truth; the Shilli-Shallas, who make statements which are alternately true and false; the Pukka-Shallas, who make two true statements followed by one false statement; and the Woppa-Shallas who make two false statements followed by one true. (Any particular set of remarks made by a member of any of the Shalla tribes could of course start anywhere in their cycle.)

Our friends and their wives were treated on the whole kindly, but it was insisted that each man should join one of the tribes (each a different one to prevent any tribe gaining an unfair advantage), and that the tribe of each man's wife should be such that, on the assumption that husbands and wives make the same number of remarks, the number of true remarks and the number of false remarks made by each married couple should be as nearly as possible equal to each other in the long run.

There were various discussions and tests to decide who should belong to which tribe. After the decisions had been made and after a certain amount of practice the five men each made three statements demonstrating their tribal truthfulness.

As follows:

ALF: (i) Queenie has 5 children.
           (ii) Priscilla is not a Wotta-Woppa.
           (iii) Ernie is not a Pukka.

BERT: (i) Duggie is a Pukka.
          (ii) My wife is a Pukka.
          (iii) Charlie is married to Sarah.
CHARLIE: (i) Tess's husband is a Pukka.
          (ii) Rachel belongs to one of the Shalla tribes.
          (iii) Bert is a Wotta-Woppa.
DUGGIE: (i) Queenie is not a Wotta-Woppa.
          (ii) Alf is a Wotta-Woppa.
          (iii) Queenie is not married to a Wotta-Woppa.
ERNIE: (i) Bert is a Pukka-Shalla.
          (ii) Charlie is a Pukka.
          (iii) Duggie is married to Rachel.

Find who is married to whom and to which tribe everyone belongs. What can you say about the number of Queenie's children?

92. More Imperfect Soccer

Soccer on the Island of Imperfection is becoming more and more popular. There are now 6 teams that compete against each other (A, B, C, D, E, F). Two of them are composed entirely of Pukkas (who always tell the truth), two of Wotta-Woppas (who never tell the truth), and the other two of Shilli-Shallas (who make statements which are alternately true and false, or false and true).

These six teams are all to play each other once.

When some of the games have been played the secretary of each team is asked the total number of matches played, the number won, lost and tied, and the number of goals scored for and against his team. The answers given will of course all be true if the team is a Pukka, all false if the team is a Wotta-Woppa, and alternately true and false, or false and true, if the team is a Shilli-Shalla.

The answers are as follows:

	Played	Won	Lost	Tied	Goals for	Goals against
A	3	3	0	0	2	0
B	3	1	2	0	7	3
C	5	2	0	3	3	4
D	4	3	1	2	3	0
E	5	0	0	5	3	2
F	5	0	4	1	0	11

An analysis of these figures showed the interesting fact that each incorrect figure was exactly 1 out, i.e., either 1 more or 1 less than the correct figure.

Find the correct figures in the table, the tribes from which A, B, C, D, E and F came, and the score in each game.

PART VIII

Assorted—Harder

93 — 102

93. The Age of Remembrance

Five middle-aged couples who had all known each other for a long time were reminiscing one evening about the predictions that some of them had made many years back.

The names of the men are Arthur, Basil, Clarence, Desmond and Ethelred; and the names of their wives, in no particular order, are Ruth, Veronica, Fanny, Polly, and Helen.

Arthur remembered that he had predicted that Clarence would marry Ruth. Basil had predicted that Ethelred would marry Veronica. Clarence had been rather less precise about his forecast and had merely said that Arthur would not marry either Polly or Helen. And Ethelred had been firmly of the view that Basil would not marry Helen.

None of the 5 marriages that would have resulted on the basis of these predictions did in fact take place.

As a matter of fact all the predictions turned out to be incorrect.

Among the ladies was Ethelred's sister. If you knew who she was you would be able to give all details about who was married to whom.

(*i*) *What 5 weddings would have resulted from the predictions?*
(*ii*) *Who was in fact married to whom and who was Ethelred's sister?*

94. Fast or Slow

We are very club conscious in my village; we have 5 altogether. The Progress, the Quicker, the Rational, the Staiput, and the Tortoise.

The Quicker and the Staiput are, as one might expect, exclusive and exhaustive (i.e., no one belongs to both, but everyone belongs to one or the other); the Progress and the Staiput are exclusive but not exhaustive; the Progress and the Rational are exhaustive but not exclusive; and the Tortoise and the Staiput are neither exclusive nor exhaustive.

(i) You are told that Smith is a member of the Progress. What can you say about his membership of other clubs?

(ii) You are told that Jones is not a member of the Quicker. What can you say about his membership of other clubs?

(iii) Adam belongs to 4 clubs. Which?

(iv) Is it possible for anyone to belong to only one club?

95. Help-the-Boys Hall

Boarding houses at the ancient, but recently modernized. school, now called Help-the-Boys Hall, are called by the initial letters of the alphabet — A, B, C, etc.; they are also called by the name of the building — The White House, The Brown House, etc.; and by the name of the housemaster.

Consider the following statements about which house a boy is in:
1. He is either in B House or the Green House.
2. He is either in A House, the White House or Mr. Smith's House.
3. He is either in D House or Mr. Jones' House.
4. He is either in the Yellow House or Mr. Budd's House.
5. He is either in C House or the Red House.

(Assume that houses referred to in the same statement are all different. For example from 2 you know that A House, the White House and Mr. Smith's House are three separate houses.)

You are told that: 1 and 2 are contradictories; 2 and 3 are contradictories; 3 and 4 are contraries; 4 and 5 are contraries.

You are also told that Mr. Codd is not the housemaster of D House, and that Mr. Budd is not the housemaster of the White House.

(Two statements are said to be contradictory if they cannot both be true and cannot both be false; two statements are said to be contrary if they cannot both be true but can both be false.)

How many houses are there? Get as much information as you can about the names and the housemasters (they have not necessarily all been mentioned) of the various houses.

96. The Light of Truth

There are three men, Smith, Jones and Brown.
    SMITH says: 'Brown is heavier than I am, and Brown is also heavier than Jones.'
    JONES says: 'I am heavier than Smith, and Smith is heavier than Brown.'
    BROWN says: 'Smith is heavier than I am, and Jones weighs the same as I do.'

Assuming that the lighter a person is the more likely he is to tell the truth, arrange Smith, Jones and Brown, in order of heaviness.

97. Anstruther and Others

Anstruther, Banks, Clopp and Dingle are, not necessarily respectively, a Journalist, a Chauffeur, a Lawyer and a Dentist.

Banks is a cousin of the Lawyer and has often stayed with him at his home in Scotland.

Clopp is 40.

The Chauffeur is three years younger than the Journalist. Anstruther is 35.

Dingle, who is a year younger than Banks, has lived all his life in Wales and has never driven a car.

The Lawyer is more than five years older than the Dentist.

Find their occupations and their ages.

98. Abracadabra Avenue

A, B, and C, live in different houses on Abracadabra Avenue, which has houses numbered from 1–80. Their numbers ascend in the order A, B, C, but none of them knows this; nor do any of them know the numbers of the houses of the other two.

They are having a conversation about it.

A thinks that B always tells the truth, and that C always lies.
B thinks that C always tells the truth, and that A always lies.
C thinks that A always tells the truth, and that B always lies.

Each one announces, not necessarily correctly, whether his number is (i) a multiple of 4, (ii) a perfect square, (iii) above 23.

A then says to B and C: 'I can tell you the numbers of your two houses, but I don't know which is which.'
B says to A: 'I can tell you the number of your house.'
C says to B: 'I can tell you the number of your house.'

They all do so, but they are all wrong; and in fact of the numbers announced, not one is the number of any of the three houses, though one of them is exactly eight times the number of one of the houses.

Where do A, B and C live?

99. What Tulsa Tortoise

Alf, Bert, Charlie, Duggie, and Ernie belong, but not necessarily respectively, to 5 different hockey teams — the Porchester Plumbers, the Queens Quicks, the Roaming Rangers, the Sadistic Saints and the Tulsa Tortoises (denoted P, Q, R, S, T). The five teams all play each other once, and after the games were over A, B, C, D and E made various remarks about the scores.

- A: (i) P vs. R was 5–2.
- (ii) R vs. T was 3–0.
- B: (i) Q vs. T was 3–1.
- (ii) Q vs. S was 0–0.
- (iii) Q vs. R was 1–1.
- C: (i) S vs. Q was 0–0.
- (ii) P vs. Q was 1–3.
- D: (i) P vs. R was 3–4.
- (ii) S vs. Q was 0–0.
- (iii) T vs. S was 4–0.
- E: (i) Q vs. R was 0–0.
- (ii) T vs. Q was 3–1.
- (iii) P vs. S was 5–1.

They were, I think, all trying to be truthful, as is indicated by the fact that any statement about a game in which the speaker played himself is perfectly correct. Unfortunately, however, every statement made about a game in which the speaker did not play is incorrect, and in every case the score of each side is one goal out — i.e., one goal more or less than the correct score.

With the additional information that P scored a total of twice as many goals as Q, that the goal averages of S and T were equal, and that no side scored more than 5 goals in one game, you should be able to discover to which team each man belonged and the score in each game.

100. The Wumbling Widgets

A certain machine is subject from time to time to three faults: the widgets wumble, the stugs stick and the sprockets fall off. The following observations are made:

When the lid is taken off, button B is pressed and the lever marked Forward is pulled back, the engine boils and the anemometer quivers.

The pressing of button B, the removal of the safety catch and a firm tap on the plate C, marked DO NOT TOUCH, are accompanied by a quivering anemometer and the input register turning blue.

If the lid is taken off, plate C tapped and the Forward lever pulled back the engine boils, and the input register turns blue.

When the safety catch is removed, plate C tapped and the Forward lever pulled back, the input register turns blue.

When the engine boils and the input register turns blue, the widgets wumble.

If the anemometer quivers and the input register turns blue, the stugs stick and the sprockets fall off.

On the assumption that the various events each have single causes, not two or more in conjunction, what would you do to be certain of curing the machine of its 3 faults?

101. 13th Avenue

Smith lives on 13th Avenue which has houses numbered from 13 to 1,300. Jones wants to know the number of Smith's house.
JONES asks: "Is it less than 500?"
Smith answers, but he lies.
JONES asks: "Is it a perfect square?"
Smith answers, but he lies.
JONES asks: "Is it a perfect cube?"
Smith answers and he tells the truth.
JONES says: "If I knew whether or not the second figure was 1, I could tell you the number of the house."
Smith tells him whether or not the second is 1 and Jones announces what he thinks is the number of the house.
But he is wrong.

What was the number of Smith's house?

102. Refined by Mogrification

(In this puzzle, unlike the others in this book, but as in all problems of decoding, it will be found that much intelligent guess work, based on a knowledge of the structure of the language, is required. The solver who takes no step unless he is certain it must be correct will take no step.)

The high priests of the tribe of Temme have discovered by long experience that prayers are only acceptable to their gods if they are first refined by mogrification. There are two stages in this process: first, transmogrification which a low priest is qualified to perform, and secondly supermogrification which can only be undertaken by the highest of the high, the Arch Priest.

The rules for the transmogrification of a sentence are as follows:
1. The longest word shall have its vowels removed and placed in the same order after the first word (thus forming the second word); its consonants shall be reversed and placed at the end of the sentence. (N.B. No sentence is mogrifiable unless it has a longest word which is not also the last word.)
2. The word that first was last shall have its consonants reversed and these shall be followed by its vowels in the original order. It shall then be placed where the longest word originally was.
3. The words that now are third and fourth shall be wedded the one to the other, and the result of this union shall be precisely bisected. (If precise bisection is not possible, then mogrification of the sentence cannot take place.) The first half shall then be reversed and placed at the beginning; the second half shall be placed at the end.

For supermogrification there is one rule only:

Words that the fingers of the priest have hitherto left untouched shall be reverse-wedded in pairs, the first with the last, the second with the last but one and so on. (If there is an odd number of hitherto untouched words the middle one shall be reversed and left in the same position.) The result of each reverse-wedding shall be precisely bisected and the first half shall be placed where the second member of the union was originally, while the second half shall be placed where the first member of the union was.

A recently discovered fully mogrified sentence ran as follows: VNRUAOHB VEASNTK UAIO IMENE SKCRTI FO ASNED RHESQIU NTRTSRF TEISFTUL.

What was the original sentence?

(*It will be found necessary to discover the rules for the wedding and reverse-wedding of words.*)

Solutions

1. Dames for Delphi

Data abbreviated:
  A: B not P.D.
  B: G to be D.O.
  G: O not H.P.
  O: I will marry Art.

If B's prediction true, then G alone predicts correctly, ∴ B's prediction false. But this is a contradiction, ∴ B's prediction not true. ∴ B not D.O., and since B's prediction is therefore not true, G not D.O.
  ∴ G's prediction false, ∴ O is H.P.
  ∴ A is D.O. (elimination). ∴ A's prediction true. ∴ B not P.D.
  ∴ B is L.W. (elimination) and G is P.D. (elimination).

Complete Solution
  Alpha becomes Delphic Oracle.
  Beta becomes Lady in Waiting.
  Omega does not marry Artaxerxes.
  Gamma becomes Professional Dancer.
  Omega becomes Harp Player.

2. Election Speculation

Set out in positive form, using obvious abbreviations, what each man says about the possibilities.
SMITH: F.F. or P.L.
BROWN: P.L. or P.P. or G.G. (*not* F.F.)
JONES: F.F. or P.P. (*not* P.L. or G.G.)

If either F.F. or P.L. or P.P. wins *two* of them are right. But if G.G. wins only Brown is right.

Since we are told that only one of them was right, the winner must have been *The Greater Glory Party*.

3. Hilarious Holidays

(i) In 1972 Whupie must be 8 Saturdays (7 + 1) after the first Saturday in February.

February 1st in 1972 is a Tuesday.

∴ first Saturday in February is February 5th.

4 Saturdays after this is March 4th.

∴ 8 Saturdays after is April 1st.

(ii) If Whupie is in February it must be not more than 3 Saturdays after first Saturday in February except in Leap Year when it might be 4 (if February 1st is a Saturday). In a Leap Year in the 1940's Whupie would be at least 5 Saturdays after the first Saturday in February.

∴ last year in which Whupie was in February was 1933.

Complete Solution
   (i) April 1st.
   (ii) 1933.

4. The Temple of Torpor

Smith is not under 30, ∴ he is not allowed to enter the T.T. of T.
  ∴ he cannot be a member of I.I.
  ∴ since he is someone of importance he is a member of Y.Y.
  ∴ he cannot be over 40.

Complete Solution
  Smith is between 30 and 40.
  He is a member of Yonkers Youngsters.
  He is not a member of Iowa Idlers.
  He may not enter the Taunton Temple of Torpor.

5. Happy Birthdays

Call them A, B, G, D, E.
  We can set out the relative positions of D and E thus: D – E.
  We know that A is as many days before G as B is after E. If B were the day after E there is no way of placing A and G in consecutive days within the limit of 5.
  ∴ B is 2 days after E, A is the day before E and G the day after.
  The order is D A E G B.

Complete Solution
  Alpha's   birthday is on Monday,
  Beta's   "   " " Thursday,
  (Gamma's   "   " " Wednesday),
  Delta's   "   " " Sunday,
  Epsilon's   "   " " Tuesday.

6. Nurses Relax

Information given, apart from that about F, can be summarized as follows:
C A,
E – D,
B – – G.
These must be combined with F to occupy 7 places. If F came between E and D, and C A between B and G, we would have E F D B C A G (or B C A G E F D, which is really the same arrangement).

But F here is not equidistant from B and C. ∴ E – D and B – – G must overlap, thus: B F E G D, or thus: E B D F G; and C A can be added at either end. In the first of these F is not equidistant from B and C, but in the second it is.

∴ E B D F G C A is the order required; and since we know that F's day off is on a Thursday we can fix the others.

Complete Solution
A's day off is on Sunday;
B's day off is on Tuesday;
C's day off is on Saturday;
D's day off is on Wednesday;
E's day off is on Monday;
F's day off is on Thursday;
G's day off is on Friday.

7. Mugs, Wumps and Others

The fact that Mugs and Wumps are contradictories can be represented thus:

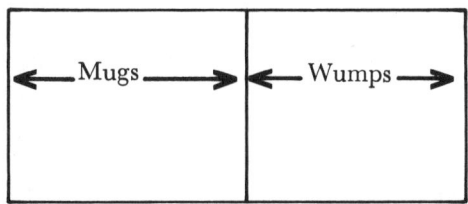

Since Wumps and Blinks are contraries, the Blinks' area must be outside the Wumps' area, and there must be an area which is neither.

Similarly the Blonks' area must be outside the Mugs' area, with an area in between which is neither.

Facts can now be represented thus:

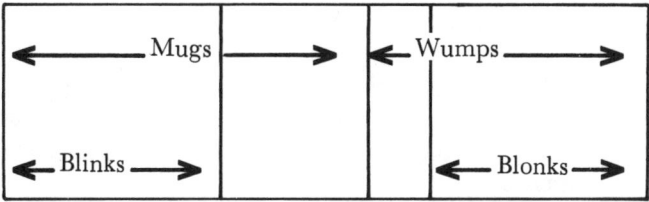

Complete Solution
    (3) and (4) are contraries.
    (a) Some Mugs are Blinks.
    (b) No Mugs are Blonks.

8. Early Examination Exits

From the fact that Smith left at 11:15 A.M. we can say that: either (1) lunch was at 1:15 P.M. and the exam started at 9:45 A.M. or later; or (2) the exam started at 9:45 A.M. and lunch was at 1:15 P.M. or later.

Similarly from Robinson's departure we can deduce that: either (3) lunch was at 1:30 P.M. and the exam started at 10:15 A.M. or earlier; or (4) the exam started at 10:15 A.M. and lunch was at 1:30 P.M. or earlier.

Either (1) or (2) *and* either (3) or (4) must both be true.

(1) and (3) cannot be true (lunch cannot be at 1:15 P.M. *and* 1:30 P.M.); similarly (2) and (4) cannot both be true.

But (2) and (3) can both be true; and (1) and (4) can both be true.

∴ *either* the exam started at *9:45* A.M. and lunch was at *1:30* P.M.; *or* the exam started at *10:15 A.M.* and lunch was at *1:15 P.M.*

9. Quietly Home to Bed

Jones fulfils the conditions in all the sentences starting 'unless' until the one 'unless their paternal grandmother does not have or did not have a passport'.

His paternal grandmother has one so we need go no further. And the fact that his surname begins with a letter in the first half of the alphabet tells us to whom he must report.

Complete Solution
   Jones should report to the Substantive Acting Registrar at 10 P.M. on Tuesday.

10. The Black and White Foreheads

Consider E's statement. If true everyone would be wearing a white disc and they would all speak the truth and say: 'I see 4 white'.

∴ *E's statement is false, and E's disc is black.*

Consider B's statement. If true there would be 1 white disc (B's) and 4 black, and all other statements false (black discs). But C's statement would be true ("I see 1 white and 3 black.").
∴ our assumption is false and B's statement is false, ∴ *B's disc is black.*

∴ A's statement is false (A sees at least 2 black). ∴ *A's disc is black.*

If C's statement false D's disc must be black (otherwise C would see 1 white and 3 black and his statement would be true). And all 5 discs would be black. But this is not possible as it would make B's statement true.

∴ *C's statement is true, and his disc is white.*

And since C sees 1 white and 3 black, ∴ *D's disc must be white.*

Complete Solution

A, B and E are wearing black discs.
C and D are wearing white discs.

11. Ditheringspoon's Spanish

(i) If A's remark false he would be 4th; but the only person to make a false remark is the one who came 5th. ∴ A's remark is true.

(ii) If B's and E's remarks both true, B would have to be 5th, and his remark false. ∴ not both true.

(iii) If B's is false he is 5th and E's is true. E and C would then have to be 1st and 4th, which makes B's true. ∴ B's is not false. ∴ E's is false (from (ii) B's and E's not both true). ∴ E is 5th.

(iv) ∴ other remarks all true. ∴ C one place above B, and D one place above A.

Since A is not 4th, this is only possible in order: D, A, C, B.

Complete Solution
1. Ditheringspoon.
2. Adams.
3. Clarke.
4. Baines.
5. Elliott.

Clarke won the Russian prize.

12. I Sit Aloof

C cannot have B or D sitting next to him, because of alphabetic condition.

∴ C has A and E next to him.

But E is not on C's right (E's brother is).

∴ A is on C's right and E on his left.

A cannot have B on his right, ∴ he must have D, and B is on E's left.

Complete Solution

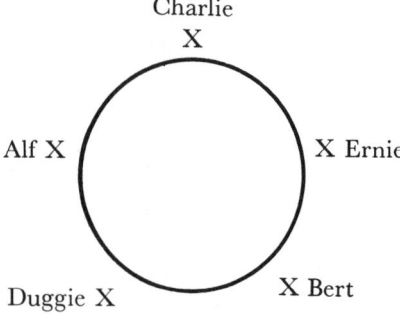

13. Just the Jobs

It will be helpful to put the data in abbreviated form:
A: E not D-K-P.
B: 1. A for W.
   2. *Not* B for W-O.
C: D for B-W.
D: B for D-K.
E: D for B-W.

(i) If C true, then C not true (because B-W), ∴ C cannot be true, ∴ C not B-W and not D-K nor D-K-P (because these are truthful).

(ii) If D true then B(1) and (2) true, then A not true, then E is D-K-P and true, then D is B-W and not true. But this is contrary to hypothesis, ∴ D not true, ∴ B not D-K, and D not D-K or D-K-P.

(iii) If A is D-K, then E not D-K-P (because A true), ∴ B is D-K-P (no one else can be), ∴ B(1) and (2) are true, ∴ A is W. But this is contrary to hypothesis, ∴ A not D-K, ∴ E is D-K (by elimination). ∴ E true, ∴ D is B-W.

(iv) Since E not D-K-P, ∴ A true, ∴ A not W, ∴ B(1) not true, ∴ B not D-K-P, ∴ A is D-K-P (elimination).

(v) If B(2) true, then B not W-O, ∴ B is W. But W's statements all false, ∴ B(2) false. But this is contrary to hypothesis, ∴ B(2) false. ∴ B is W-O, ∴ C must be W.

Complete Solution

Alf	Door-Knob-Polisher
Bert	Welfare Officer
Charlie	Worker
Duggie	Bottle-Washer
Ernie	Door-Keeper

14. Body and Soul

1. Consider C. If his remark is true it cannot be true, for only the lightest man's statement is true.
∴ *C false*, and C is lighter than D.

2. Since D is heavier than C, ∴ D is false, and C is lighter than B.

3. ∴ B is not the lightest, ∴ by elimination *A is the lightest* and *A's remark is true*.

4. ∴ B is lighter than D, and from (1) C is lighter than D, and from (2) C is lighter than B. ∴ C is second lightest; and B is third.

Complete Solution
(Lightest first)   1. A.
                   2. C.
                   3. B.
                   4. D.

15. Who Works When?

1st day: A C E
2nd day: B E   (Conditions for B are: C present, D absent; and for E: A present and B absent. These are fulfilled on the 1st day, but not the conditions for any of the others.)
3rd day: A D
4th day: C E
5th day: B D
6th day: A C
7th day: B E

The 7th day is the same as the 2nd, but this is the first repetition. Clearly the cycle BE, AD, CE, BD, AC, BE etc., will repeat itself indefinitely. Note that there are 5 different arrangements repeating themselves so that BE will recur on the 2nd, 7th, 12th, 17th, 22nd etc. days.

The 100th day will be the same as the 5th day.
The 383rd "   "   "   "   "   "   " 3rd day.

Complete Solution
On the 100th day Bert and Duggie will be present.
On the 383rd day Alf and Duggie will be present.

16. Chagrin for Charlie

We know (i) that C was not 1st or 5th; (ii) that E was not 2nd and not 4th or 5th (because D was 2 places lower); (iii) that D was not 1st, 2nd or 4th (he was 2 places lower than E who was not 2nd); (iv) that A was not 1st or 5th; (v) that B was not 1st or 2nd (he was 1 place below C who was not 1st).

This information can be put in a diagram, thus:
1. E
2. C   A
3. C E D A B
4. C   A B
5. D   B

The only candidate for the 1st place is E. And since D was 2 places below E, ∴ D was 3rd. B was 1 place below C and the only vacant adjacent places are 4th and 5th. ∴ C was 4th and B 5th. ∴ A must have been 2nd.

Complete Solution
1. E.
2. A.
3. D.
4. C.
5. B.

17. When Are We?

Call them, A, B, C, D, E, F, G. Put in positive form which day each man says it is.
  A: Monday.
  B: Wednesday.
  C: Tuesday.
  D: Thursday, Friday, Saturday or Sunday.
  E: Friday.
  F: Wednesday.
  G: Monday, Tuesday, Wednesday, Thursday, Friday or Saturday.

The only day that is mentioned only once is Sunday. If it were any other day of the week more than one of these remarks would be true.

Therefore it must be *Sunday*, and Duggie's remark is true.

18. Working to Rules

(i) From (1), with A present the arrangement must either be:

	Present	Absent
(a)	A	B
or (b)	A, B	C, E

(remaining letters to be filled in later).

(ii) From (2), C absent in (a). From (1) and (3) E present and D absent in (a). In (b) D may be present or absent, so that we now have 3 possibilities:

	Present	Absent
(a)	A, E	B, C, D
(b)	A, B, D	C, E
(c)	A, B	C, D, E

(and these are the only possibilities with A present).

(iii) From (4) another arrangement is:

	Present	Absent
	C, D	B, E

and from (2) A must be absent, ∴ we have:

	Present	Absent
(d)	C, D	A, B, E

(iv) Consider other arrangements with A absent and ∴, from (2), C present.

	Present	Absent
	C	A

Suppose E present, and ∴, from (3), D absent

	Present	Absent
	C, E	A, D

From (4) we cannot have B absent, but we can have B present; giving another arrangement:

	Present	Absent
(e)	B, C, E	A, D

(v) Suppose now, with C present and A absent, we have E absent.

	Present	Absent
	C	A, E

If we have B absent then, from (4), we get D present, and thus (d) again. ∴ suppose B present, then D may be present or absent, giving:

	Present	Absent
(f)	B, C, D	A, E
or (g)	B, C	A, D, E

We have now exhausted all the possibilities.

Complete Solution

The seven arrangements are:

Present	Absent
A, E	B, C, D
A, B, D	C, E
A, B	C, D, E
C, D	A, B, E
B, C, E	A, D
B, C, D	A, E
B, C	A, D, E

19. Who Lives Where?

Suppose that A's number is x. If all the statements were true the situation would be:

(i) A: x.
(ii) B: x − 23.
(iii) C: x − 7.
(iv) D: x + 12.
(v) E: x.
(vi) F: x − 30.
(vii) G: x − 13.
(viii) H: x + 24.
(ix) J: x + 12.

But it is not possible for A and E to have the same numbers.
∴ one of the statements made by A, B, C and D is false.
Again since the numbers above for D and J are the same;
∴ one of the statements made by D, E, F, G, H is false.
But there is only *one* false statement, ∴ *it must be D's.*
And we know that all the other statements are *true*.
D incorrectly gives E's number as x (the same as A's).
Let us call E's number y. Then we have:

E: y.
F: y − 30.
G: y − 13.
H: y + 24.
J: y + 12.

But there are only 55 houses.
∴ we must have y = 31, and numbers are, E 31, F 1, G 18, H 55, J 43.

J's number is 10 more than A's, which is therefore 33. And since we know that A's, B's and C's statements are true,
∴ B's number is 10, C's is 26, and D's is 45.

Complete Solution

Alf lives at number 33.
Bert " " " 10.
Charlie " " " 26.
Duggie " " " 45.
Ernie " " " 31.
Fred " " " 1.
George " " " 18.
Hubert " " " 55.
Judith " " " 43.

20. Sammy the Soothsayer

A diagram will help, in which information can be inserted as it is discovered.

Numbers		Present				Future			
		B-W	D-O	D-S	W	B-W	D-O	D-S	W
	A	×				×			
15	B	×				×			
11	C		×	×	×	×	×		
27	D	×	×			×		×	×
	numbers						27	even	

    (i) (5), (8) and (11) cannot all be true. This gives us at least *one* false remark.

    (ii) (3), (7) and (12) cannot all be true. (The future B-W and the present B-W cannot have the same numbers). This gives us at least *one more* false remark.

    (iii) (1) and (6) cannot both be true. This gives us a 3rd false remark. And since there are only 3 false remarks, ∴ all the other remarks ((2), (4), (9), (10), and (13)) are true.

(The facts given by the truth of (2), (4) and (9) have been inserted in diagram.)

    (iv) We now know that one and only one of (5), (8) and (11) is false. (9) is known to be true and it contradicts (5). ∴ *(5) is false, and (8) and (11) are true.*

    (v) Since (2) is true, ∴ C is B-W, ∴ *(7) is false,* ∴ *(3) and (12) are true* (we know that one and only one of (3), (7) and (12) is false).

(Further facts should be inserted in diagram, remembering

that when we know that D is future D-O we know also that D is *not* the present D-O.)

(vi) We now know that B, C and D's numbers are all odd. And from (4) (true) we know that future D-S number is even, ∴ *A must be future D-S.* (∴ A is not present D-S.) *And by elimination C is future W and B is future B-W.*

And since we know that the number of C (future W) is 11 ∴ (1) is true and (6) is false.

(vii) *Consider (13).* (*True*). Suppose present W is D, then D's number (27) is greater by 1 than B's (15) + C's (11) + A's (?). But even if A's number were 1, this is not possible.

∴ present W is not D, and in order that (13) should be true *present W can only be A* and ∴ *A's number is 54* (15 + 11 + 27 + 1).

(viii) ∴ by elimination *D is present D-S*, and *B is present D-O*.

Complete Solution

	Present job	*Future job*	*Number*
Alf	Worker	Door-Shutter	54
Bert	Door-Opener	Bottle-Washer	15
Charlie	Bottle-Washer	Worker	11
Duggie	Door-Shutter	Door-Opener	27

21. Muscles Are Not Enough

From the bottom line, total of marks scored = 105. ∴ D must have scored 20.

Consider question 4. Only 1 person can have got it right (total points 15). ∴ answer must be 20% and A got it right.

Consider question 3. Total of 35 must be made up of three 10's and one 5 or two 10's and three 5's. But since there are not three answers the same it must be two 10's and three 5's. ∴ 21 is the answer and the three wrong answers each received 5 points.

∴ E got 10 points for question 3; but only 10 points altogether. ∴ 0 for all the others, and other answers wrong.

∴ 4 not right answer for question 1. ∴ it must be 3 (total of 30 not otherwise attainable), and B and C must each have got 5.

B only got 15 altogether, ∴ B got 0 for questions 2, 4 and 5.

Only one person got question 2 right. Not B or E. And not A or D for their answers are the same. ∴ C got it right. ∴ answer is 18° 30'. A and D both got 0.

By subtraction A got 5 for question 5.

C or D must have got question 5 right. But not D for that would make his total too big. ∴ C, and answer is $143.

∴ C got 0 for question 4. ∴ D got 5 (total for question is 15), and D got 0 for question 5.

Complete Solution

Points for each question

	1	2	3	4	5
Alf	10	0	5	10	5
Bert	5	0	10	0	0
Charlie	5	10	5	0	10
Duggie	10	0	5	5	0
Ernie	0	0	10	0	0
Correct answers	3	18° 30'	21	20%	$143

22. Wealth, Happiness and Health

Convenient to have the data in more compact form, using obvious abbreviations.
  A: 1. E's nurse is T.
  B: 1. C's nurse is S.
     2. D's room number is 102.
     3. A's room number not greater than 100.
  C: 1. D's room number is a prime.
  D: 1. A's nurse is S.
  E: 1. B's room number is a multiple of 7.

The two in the highest numbered rooms are truthful; from the data their nurses must be S and T. The nurses of the other three (who are not truthful) in lower-numbered rooms are P, Q and R.

From the information we have, the 3 untruthful invalids cannot have room numbers higher than 101 (leaving 102 and 103 for the truthful); and the two truthful ones cannot have room numbers less than 100 (leaving 97, 98, and 99 for the untruthful).

A diagram will help. Thus:

True or false	P	Q	R	S	T	Room numbers
A						101, 102 or 103
× B				×	×	
C				×		
× D				×	×	not 102
E						

1. If B(2) true, then D is truthful; if D(1) true, A is truthful. But we are told that only 2 are truthful. ∴ B(2) *cannot be true*. ∴ *all B's remarks are false* (this fact has been marked in diagram).

2. ∴ C's nurse not S (mark in diagram as shown) and A's room number is greater than 100. (This information has been inserted in diagram; note that it does not tell us whether A is truthful.) And D's room number is not 102 (insert in diagram).

3. If D(1) true, then A(1) true; and if A(1) true then E(1) true. But we know that only 2 of the patients are truthful, ∴ D(1) *cannot be true*. (Indicate in diagram as shown.) And since B and D are both untruthful, ∴ their nurses cannot be S or T (insert in diagram as shown; other facts should be inserted as they are discovered).

4. Since D(1) not true, ∴ A's nurse not S. ∴ by elimination *E's nurse is S*.

5. ∴ E is truthful, ∴ *B's room number must be 98* (the only multiple of 7). Also A(1) not true (E's nurse is S, not T). ∴ C must be other truthful one. ∴ *C's nurse is T*.

6. Since A's room number greater than 100 (see 2) and less than 102 (since A is a liar). ∴ *A's room number is 101*, and room numbers of S and T must be 102 and 103 in that order.

7. Since A's room number is 101, this must be highest of P, Q, R, ∴ *A's nurse is R*.

8. From C(1) we know that D's room number is a prime; we also know that it must be less than 101 (R is in 101). ∴ *it must be 97*, and *D's nurse must be P*. ∴ *B's nurse is Q*.

Complete Solution

Alf's	nurse is	Rachel;	his room number is			101.
Bert's	"	" Queenie;	"	"	"	" 98.
Charlie's	"	" Tess;	"	"	"	" 103.
Duggie's	"	" Priscilla;	"	"	"	" 97.
Ernie's	"	" Sarah;	"	"	"	" 102.

23. Mr. Bottles and Mr. Doors Go to the Convention

Call the 4 speakers A, B, C and D in that order.

(i) *Consider $A(1)$, $C(2)$ and $D(2)$.* At least two of these must be false, ∴ not more than one of Bert, Charlie and Fred present. And A is not Fred, C is not Bert, D is not Charlie. (It will be convenient to mark these and other facts as they are discovered in a diagram.)

(ii) *Consider $A(2)$, $B(1)$ and $D(1)$.* At least two of these are false, ∴ not more than one of Alf, Bert and Duggie present. And A is not Bert, B is not Alf, D is not Duggie.

(iii) From (i) at least two of Bert, Charlie and Fred absent; from (ii) at least two of Alf, Bert and Duggie absent. But only 3 absent altogether; ∴ *Bert must be absent;* and either Charlie or Fred is absent; and either Alf or Duggie is absent; and the other two, Ernie and George, are present. Also we know that either Charlie or Fred (whichever is present) is B-W; and either Alf or Duggie (whichever is present) is Worker.

(iv) *Consider $C(1)$.* We know that Ernie is present; ∴ *this is true.*

(v) *Consider $B(2)$.* (Charlie is not C.) If false Charlie is C, and is present, but what is said about those present is true. ∴ This cannot be false; Charlie is not C, but he is present. We know that Charlie is not B or D, as they make remarks about him; ∴ *Charlie is A.* ∴ Charlie is present and is B-W, and Fred is absent.

(vi) *Consider $A(3)$.* (Duggie is not C.) By the same argument as in (v) this is true, ∴ Duggie is present but is not C. ∴ by elimination *Duggie is B.* ∴ Duggie is present and is Worker, and Alf is absent.

(vii) We know that Ernie is present (see (iii)), and is W-O. He cannot be A, B or C (C mentions him). ∴ *Ernie is D.* ∴ *George is C;* and from $D(3)$ (true) *George must be Sweeper-Upper.*

Complete Solution
    Mr. Welfare is Charlie and is Bottle-Washer.
    Mr. Sweepers is Duggie and is Worker.
    Mr. Bottles is George and is Sweeper-Upper.
    Mr. Doors is Ernie and is Welfare Officer.

24. How Faithful Our Bottle-Washer

D(2) is a starting point. If true, D never tells truth, ∴ false, ∴ D could not have been there for more than $2\frac{1}{2}$ days, ∴ D was there for $2\frac{1}{2}$ days exactly, ∴ on Wed., and D(1) true. ∴ F there for 4 days, ∴ F(1) and (2) both true, ∴ E(1) and (2) both true. From E(2) since E is truthful and was therefore there for more than $2\frac{1}{2}$ days, E was there on Monday, Thursday and Friday. ∴ from E(1) A was there on Tuesday and Wednesday, ∴ A's statements are both false. ∴ G is not B-W and since neither A, D, E nor F are always present none of them are B-W. ∴ B or C is B-W. But if B were B-W, B(1) would be true. But B-W must be there on Wed. ∴ B not B-W ∴ C is B-W and both his statements are true, and C is there every day.

G(2) is false (we know that only one of D's statements is true). If G(1) true, then G present $2\frac{1}{2}$ days, since his statements are alternately true and false, ∴ G present on Wed. But this contradicts G(1), ∴ G(1) false. ∴ G was present on Wed., and on only one other day.

Consider rule (iii). A present on Tues. and D not absent on Wed. ∴ B present on Wed.

A also present on Wed., but D absent on Thurs.
∴ B and G absent on Thurs.

We now know that B(1) false, and since C(1) is true, ∴ B(2) is true. ∴ G present on Tues. ∴ from rule (ii) B not present on Tues., and we know B not present on Thurs., ∴ B present on Mon., Wed., Fri.

Complete Solution
  Monday: B, C, D, E, F;
  Tuesday: A, C, D, F, G;
  Wednesday: A, B, C, D, G;
  Thursday: C, E, F;
  Friday: B, C, E, F.
  Charlie is the Bottle-Washer.

25. We Polish Up Our Percentages

A table will help, thus:

	D-O	D-S	D-K-P	B-W	S-U	W-O	W
A	✓ ✗	✓ ✗	✗	✓	✗	✓	✓
B					✗		
C							
D							
E			✗				
F			✗				
G			✗	✗	✗	✗	✗

This reminds us who tells the truth, etc., and enables us to insert in a compact form the information which we obtain.

Consider G(1). If false G is W, but W always tells the truth, ∴ G(1) cannot be false, ∴ G not W, and G(1) true. If G(2) true, B's statements both false, ∴ B is D-O, ∴ one of B's statements true and one false, ∴ G(2) not true, ∴ G's statements alternately true and false, ∴ G either D-O or D-S. And since G(3) must be true D-K-P is not E or F. (Information obtained so far has been inserted in diagram.)

Since G is either D-O or D-S, ∴ C(1) false, ∴ F(1) false, ∴ B(1) false, ∴ neither B, C nor F can be W-O, W or B-W. If B(2) false then B is D-O, but B(1) also false, ∴ B cannot be D-O, ∴ B(2) true. ∴ *B is D-S* (the only other man who makes remarks which are alternately true and false). ∴ by elimination G is D-O, and F is S-U. ∴ F(2) is false. A, D and E are now the only 3 who have not told at least one lie, ∴ they must be the B-W, the W-O and the W, but we do not yet know which is which. All their statements are true. And C must be the D-K-P and C(2) is false. From D(1) D is not the W-O.

Consider wages. From A(1), A:B = 6:5; from A(2), B:E = 21:20; from E(1), A's new wage is 27/25 ($\frac{108}{100}$) of A's old wage; ∴ A's wage is a multiple of 6, of 21 and of 27; ∴ it must be 378 dollars (378 is the only multiple of these three numbers which is less than 400). ∴ B's wage is 315 dollars and E's is 300 dollars. From G(3) D-K-P (C) gets 305 dollars and F (S-U) gets 315 dollars. From D(1) W-O's wage must be a multiple of 4; but we know that W-O is either A (378) or E (300). ∴ W-O is E. And since D gets 10 dollars more than W-O (from D(1)), ∴ D gets 310 dollars. From E(2) W's wage must be a multiple of 5, ∴ W cannot be A (wage 378 dollars) and must be D (wage 310 dollars). ∴ by elimination A is B-W. And D-O (G's) wage is 248 dollars (20% less than 310).

Complete Solution

Alf	Bottle-Washer	$378	per month
Bert	Door-Shutter	315	" "
Charlie	Door-Knob-Polisher	305	" "
Duggie	Worker	310	" "
Ernie	Welfare Officer	300	" "
Fred	Sweeper-Upper	315	" "
George	Door-Opener	248	" "

26. How Vocal Our Voters

A table will help.

	D-K-P	D-O	D-S	S-U	W-O	B-W	W	Political party
A								
B								
C					X			Dem.
D								
E								Soc.
F					X			Pr.
G								
Political Party				Soc.				

(i) If E(1) true, E Dem. But Dems start with false statement, ∴ E(1) not true, ∴ E not Dem, ∴ *E Soc.* (None of the others start with a false statement.) Similarly C(1) cannot be true, ∴ *C Dem.* Similarly F(1) cannot be false, ∴ *F Pr.*

(ii) Since C Dem., ∴ C(3) false, ∴ W-O Soc. ∴ W-O not C or F.

(Facts discovered so far have been inserted in diagram; it will be helpful to insert more facts as they are discovered.)

(iii) Since F Pr., ∴ F(2) and F(3) true, ∴ S-U Dem., ∴ S-U not E or F; and A not W. Since E Soc., ∴ E(3) false, ∴ A not D-K-P.

(iv) If B(2) true, G(1) true, and ∴ B(2) false. But this is a contradiction, ∴ B(2) false. ∴ B is Soc., or Rep. (2nd remark false). ∴ B(1) true. ∴ *B Rep.* ∴ B not S-U or W-O; and G not Pr. (B(2) false).

(v) Since B Rep., ∴ D(2) false, ∴ D(1) false (Prs. always truthful), ∴ D Soc., ∴ D not S-U.

(vi) G(2) true, ∴ G(1) false, ∴ *G Dem.*, ∴ G not W-O.

(vii) C(2) true (Dem.'s 2nd remark). ∴ *B-W Soc.;* B-W not B, C, F, G.

(viii) E(2) false, ∴ *D-O Dem.*, ∴ D-O not B, D, E, F.

(ix) If A Rep., A(1) true. But this is a contradiction, ∴ A

not Rep., ∴ A(2) true (we can see that B is the only Rep). ∴ *A Pr.* ∴ A not D-O, S-U, W-O, B-W. ∴ by elimination *A is D-S*.

(x) F(3) true, ∴ W not Pr., ∴ W not F. ∴ by elimination F is *D-K-P*. ∴ by elimination *B is W*.

(xi) From C(2), D not B-W, ∴ by elimination *D is W-O*, and *E is B-W*. From E(2), G not D-O, ∴ by elimination *G is S-U*, and *C is D-O*.

Complete Solution

Alf	Door-Shutter	Progressive
Bert	Worker	Republican
Charlie	Door-Opener	Democrat
Duggie	Welfare Officer	Socialist
Ernie	Bottle-Washer	Socialist
Fred	Door-Knob-Polisher	Progressive
George	Sweeper-Upper	Democrat

27. The Pay Roll Rules

(i) From 2, E gets $\frac{112}{100} \times \frac{110}{100}$ (i.e. $\frac{154}{125}$) of B-W's present wage. This gives us the ratio of E's wage to the B-W. Since they must be whole numbers of dollars, between $200 and $600 (from 5), and since B-W's must be divisible by 10 (he gets a 10 per cent raise next month), ∴ E's wage = $308 and B-W's wage = $250.

(ii) From 6, D-K gets $\frac{105}{100} \times \frac{90}{100}$ (i.e. $\frac{189}{200}$) of B's wage. Since no wages are less than $200 or greater than $600, ∴ D-K's wage must be $378 or $567 and B's wage must be $400 or $600. And E not B-W or D-K; and B not D-K or B-W.

(iii) From 3, D-K-P gets $\frac{130}{100}$ (i.e. $\frac{13}{10}$) of old wage, ∴ wage is divisible by 13; ∴ D-K-P not B or E.

(iv) From 4, C gets $(\frac{120}{100}x - 12)$ dollars (where x is W-O's wage), i.e. C gets $(\frac{6}{5}x - 12)$ dollars. ∴ W-O's wage is div. by 5; ∴ W-O not E; ∴ by elimination *E is W* and *B is W-O*.

(v) B's wage is $400 or $600 (see (ii)). If $600, C's wage would be $(\frac{6}{5} \times 600 - 12)$ which is more than $600. ∴ *B's wage is $400*. ∴ C's wage is $(\frac{6}{5} \times 400 - 12)$ dollars, i.e. $468. ∴ from (ii) D-K's wage is 378 dollars. ∴ C not D-K. From (i) B-W's wage is $250, ∴ C not B-W; ∴ by elimination *C is D-K-P*.

(vi) ∴ A and D are between them D-K ($378) and B-W ($250), ∴ from 1, *A is D-K* and *D is B-W*.

Complete Solution

Alf	Door-Keeper	$378	per month
Bert	Welfare Officer	400	" "
Charlie	Door-Knob-Polisher	468	" "
Duggie	Bottle-Washer	250	" "
Ernie	Worker	308	" "

28. The Higher the Truer

A table will help:

	Room number	Temperature	Pajamas
A			
B		103°	
C		97°	
D		100°	
E		99°–102°	Pink
F		101°	
G		98°	Pink

(i) Consider E(2). If true, E(1) would have to be true also. But if both true E's remarks would have to be alternately true and false, which they are not. ∴ E(2) false, and since E's temperature is not under 99°, ∴ E(1) and E(3) both true. (∴ E's pajamas are pink.) And since E makes a false remark ∴ E's temperature is between 99° and 102°.

(ii) Consider G(3). If true C always tells truth. ∴ from C(1) F's temperature is 98°, and as, from C(3), he is wearing blue pajamas F never tells truth. ∴ from F(3), C's temperature not over 102°. As this contradicts original hypothesis, ∴ G(3) is false. ∴ F(3) is false, ∴ F(1) is false. ∴ F's temperature is 101°, ∴ F(2) and F(4) are true.

(iii) Since E(3) is true, ∴ B's temperature is 103°, ∴ B(1), (2), (3) all true. ∴ D's temperature 100° and ∴ G's temperature 98°, and since G(3) false, ∴ G(1) false, ∴ G's pajamas pink, ∴ G(2) and (4) true. ∴ C's temperature 97°.

(Facts discovered so far have been inserted in diagram. The reader is advised to insert further facts in his own diagram.)

(iv) From B(2), D's statements alternately true and false. D(2) is false (D's temperature is 100°, and from F(2) his room number must be 101 or 103). ∴ D(1) and D(3) true, and D(4) false. ∴ A's pajamas green.

(v) F(2) true, ∴ A(3) is true. ∴ A's temperature not under 99° (he cannot be wearing pink pajamas as E and G are wearing the only two pairs), ∴ from D(3) (true) A's room number at least 100. And since we know that F's temperature is 101°, ∴ from B(1) (true) A's room number is 104, and from D(3), A's temperature is 103°. ∴ A(1), (2) and (3) all true. ∴ F's pajamas are aquamarine and C's room number is 99.

(vi) Consider C. We know that his temperature is 97°, and that he is not wearing pink pajamas (E and G are, but no one else). ∴ all C's statements are false. ∴ from C(4), C is wearing yellow pajamas.

(vii) We now know the colors of all pajamas except B's and D's. But we know that D is not wearing flame (D(4) false), ∴ B is wearing flame and D blue.

(viii) D's room number is 101 or 103 (see (iv)). *Suppose 101:* since C's room is 99 and A's is 104, the only two consecutive numbers for B's and E's rooms (F(4) true) would then be 102 and 103. It would not then be possible for F's room number to be 2 greater than B's (G(4) true). ∴ D's room number must be 103.

(ix) Since B's room is next to E's and E's number is odd (C(2) false), and F's is 2 greater than B's,

∴ F must be in 102,
    E " " " 101,
and B " " " 100.
∴ by elimination G must be in 98.

Complete Solution

	Room number	Temperature	Pajamas
Alf	104	103°	Green
Bert	100	103°	Flame
Charlie	99	97°	Yellow
Duggie	103	100°	Blue
Ernie	101	99°–102°	Pink
Fred	102	101°	Aquamarine
George	98	98°	Pink

29. Who Are We?

Numbers of remarks		D-K-P	D-O	D-S	B-W	S-U	W-O	W	Wives A	B	C	D	E	F	G
	Alf								×						
	Bert									×					
4, 13, 16	Charlie	×	×								×				
	Duggie											×			
	Ernie												×		
	Fred													×	
	George														×

		D-K-P	D-O	D-S	B-W	S-U	W-O	W
	A	×						
	B	×						
	C	✓	×	×	×	×	×	×
Wives	D	×						
	E	×						
	F	×						
	G	×						

Remarks

True or false and by whom:
1. Ernie is W.
2. Fred is not D-O.
3. Gertie is married to W-O.
✓ C 4. Clarissa is married to D-K-P.
5. Bert is not the D-O.
6. George is married to Diana.
7. Agnes is not married to Bert.
8. Duggie is the W.
9. Charlie's wife is Flossie.
10. Clarissa is married to Bert.
11. Duggie is married to Beatrice.
12. Alf is the B-W.
× C 13. Charlie is the D-O.
14. George is the D-K-P.
15. George is not the D-S.
× C 16. Charlie is the D-S.
17. Ernie is the W-O.

18. Ernie is married to Gertie.
19. The numbers of 2 of Bert's remarks are perfect squares.
20. Charlie is the B-W.
21. Fred is the D-O.

(i) Insert in the diagram the fact that Alf is not married to Agnes, etc. (as shown).

(ii) Total of numbers of remarks = $1 + 2 + 3 + \ldots + 21 = 231$. $231 \div 7 = 33$. ∴ each man's total is 33 except for Duggie and Fred, one of whom must have a total of 32 and the other 34.

(iii) Consider (13), (16), (20). At least 2 of these must be false, ∴ at least 2 must be by Charlie. These 2 must be (13) and (16) as otherwise total greater than 33. ∴ Charlie's 3rd remark must be (4), which is true. (These facts have been inserted in diagram. The reader is recommended to insert others in his own diagram as they are obtained.)

(iv) (20) is not by Charlie and is therefore true. (∴ Charlie is B-W.) (9) is not by Charlie and is therefore true. (∴ Charlie's wife is Flossie, and Flossie is married to B-W.)

(v) (3) is true because no man's name comes into it. (∴ Gertie is married to W-O.) ∴ (17) and (18) are both true or both false. If both false, both made by Ernie, which is impossible ($17 + 18 = 35$), ∴ both true. ∴ (1) is false, ∴ (1) by Ernie. (12) is false, ∴ (12) is by Alf.

(vi) Ernie's other 2 remarks (besides 1) must be (21) and (11), or (20) and (12), or (19) and (13), or (18) and (14), or (17) and (15), to add up to 33. But (12), (13), (17), (18) are all known not to be by Ernie. ∴ (21) and (11) are by Ernie, ∴ both true (∴ Duggie is married to Beatrice, and Fred is D-O.)

(vii) (2) is false, ∴ by Fred. (5) is true, ∴ not by Bert. (19) is false (we know that 1, 4, 16 are not by Bert), ∴ it is by Bert. ∴ Bert's other two remarks add up to 14. A look at those which we know to be by someone else shows that they can only be (6) and (8). ∴ (6) is true (George is married to Diana) and (8) is true (Duggie is W). And as Duggie is married to Beatrice, ∴ Beatrice is married to W.

(viii) George is married to Diana, and Diana not married to D-K-P (Clarissa is), ∴ George not D-K-P, ∴ (14) is by George.

(ix) Since (10) is not by Bert, ∴ true. (Clarissa is married to Bert.) And since Clarissa is married to D-K-P, ∴ Bert is D-K-P.

(x) By elimination Alf is married to Ethel, and Fred is married to Agnes.

(xi) (14) is by George, ∴ his other 2 remarks add up to 19. Eliminating those remarks which we know to be by someone else these can only be (9) and (10). ∴ (15) not by George, ∴ true (∴ George is not the D-S). ∴ by elimination George is S-U and Alf is D-S. The diagram giving names, jobs and wives can now be completed.

(xii) (12) is by Alf. ∴ his other two add up to 21. They can only be (3) and (18). (2) is by Fred and remarks still unadopted are (5), (7), (15), (17), (20). It is easy to see that Fred's must have been (2), (15) and (17); and Duggie's were (5), (7) and (20).

Complete Solution

Alf is the Door-Shutter, is married to Ethel and makes remarks (3), (12), (18).

Bert is the Door-Knob-Polisher, is married to Clarissa, and makes remarks (6), (8), (19).

Charlie is the Bottle-Washer, is married to Flossie, and makes remarks (4), (13), (16).

Duggie is the Worker, is married to Beatrice, and makes remarks (5), (7), (20).

Ernie is the Welfare Officer, is married to Gertie, and makes remarks (1), (11), (21).

Fred is the Door-Opener, is married to Agnes, and makes remarks (2), (15), (17).

George is the Sweeper-Upper, is married to Diana and makes remarks (9), (10), (14).

30. The Sad Demise of Derrick Demmit

ALF: In P at 2:32 P.M.; in S at 4:54 P.M.
 (i) *To A.A.* Helicopter at 2:34 P.M.; arrive at 3:04 P.M.
 (ii) *From A.A.* Car; leaving at 3:06 P.M.
 ∴ *in A.A. from 3:04 P.M. to 3:06 P.M.* ∴ *not guilty.*

BERT: In Q at 1:47 P.M.; in S at 4:56 P.M.
 (*not* by helicopter)
 (i) *To A.A.* Car; arriving at 2:59 P.M.
 (ii) *From A.A.* Car; leaving at 3:08 P.M.
 ∴ *in A.A. from 2:59 P.M. to 3:08 P.M.* ∴ *not guilty.*

CHARLIE: In R at 2:09 P.M.; in P at 4:02 P.M.
 (i) *To A.A.* Helicopter at 2:10 P.M.; arriving at 3:10 P.M.
 (ii) *From A.A.* Car; leaving at 3:14 P.M.
 ∴ *in A.A. from 3:10 P.M. to 3:14 P.M.* ∴ *could be guilty.*

DUGGIE: In S at 1:25 P.M.; in P at 3:40 P.M.
 (i) *To A.A.* Train at 1:26 P.M.; arriving at 3:04 P.M.
 (ii) *From A.A.* Helicopter, leaving at 3:09 P.M.
 ∴ *in A.A. from 3:04 P.M. to 3:09 P.M.* ∴ *not guilty.*

ERNIE: In S at 1:56 P.M.; in R at 4:41 P.M.
 (i) *To A.A.* Helicopter at 1:58 P.M.; arriving at $3:05\frac{1}{2}$ P.M.
 (ii) *From A.A.* 3:08 P.M. train.
 ∴ *in A.A. from $3:05\frac{1}{2}$ P.M. to 3:08 P.M.* ∴ *not guilty.*

FRED: In Q at 1:52 P.M.; in P at 3:59 P.M.
 (i) *To A.A.* Car; arriving at 3:04 P.M.
 (ii) *From A.A.* 3:12 P.M. train.
 ∴ *in A.A. from 3:04 P.M. to 3:12 P.M.* ∴ *could be guilty.*

GEORGE: In R at 2:11 P.M.; in Q at 4:37 P.M.
 (i) *To A.A.* Train; arriving at 3:44 P.M.

(ii) *From A.A.* Helicopter, leaving at 3:50 P.M.
∴ *in A.A. from 3:47 P.M. to 3:50 P.M.* ∴ *not guilty.*

Complete Solution

Alf's earliest time at A.A. was 3:04 P.M.; his latest time was 3:06 P.M.

Bert's earliest time at A.A. was 2:59 P.M.; his latest time was 3:08 P.M.

Charlie's earliest time at A.A. was 3:10 P.M.; his latest time was 3:14 P.M.

Duggie's earliest time at A.A. was 3:04 P.M.; his latest time was 3:09 P.M.

Ernie's earliest time at A.A. was $3:05\frac{1}{2}$ P.M.; his latest time was 3:08 P.M.

Fred's earliest time at A.A. was 3:04 P.M.; his latest time was 3:12 P.M.

George's earliest time at A.A. was 3:44 P.M.; his latest time was 3:50 P.M.

Either Charlie or Fred could be guilty.

31. 3 Teams

Since A won both games therefore B and C both lost against A. And since B tied a game therefore B tied C.

The total of all goals for must equal the total of all goals against.

Therefore A scored 7 goals. B and C between them scored 5 goals, one of which was scored against A. Therefore they scored 4 goals against each other in their tied game. Therefore the score was 2–2. By simple subtraction we can now deduce that B lost to A by 0–2, and that C lost to A by 1–5.

Complete Solution

	Played	Won	Lost	Tied	Goals for	Goals against
A	2	2	0	0	7	1
B	2	0	1	1	2	4
C	2	0	1	1	3	7

Scores: A vs. B   2–0,
          B vs. C   2–2,
          A vs. C   5–1.

32. 5 Teams

A table will be useful, thus:

	A	B	C	D	E
A	×	×	T 0–0	W	W
B	×	×			
C	T 0–0		×	T	
D	L		T	×	
E	L				×

Information as to who played whom, and the results and scores can be inserted as they are discovered. And it will also be useful to fill in gaps in the diagram that is given in the question.

(i) The total of games lost must be equal to total won; ∴ E lost 3.

(ii) Total goals for = total goals against. ∴ E had 8 goals against.

(iii) By subtracting number of games won and lost from total we see that A tied 1, C tied 2, D tied 1. ∴ A vs. C and C vs. D must both have tied. And since A had no goals scored against ∴ score in A vs. C was 0–0.

(iv) A could not have played B, who won both their matches, ∴ A played and beat D and E.

(These facts have been inserted in diagram; others should be added as they are discovered.)

(v) E could not have played C (both E and C lost all their games). ∴ E played, and lost to, B and D.

(vi) C's other match must have been against B (lost) and since B only played 2 matches B did not play D. We now know who played whom and the result of each match.

(vii) B had 1 goal scored against, not by E (who scored none), ∴ by C. And since C scored 2 goals, but none against

A, ∴ C scored 1 against D, and score was 1–1. ∴ by subtraction C vs. B was 1–3.

(viii) Consider B. Since B vs. C was 3–1, ∴ B vs. E was 1–0.

(ix) D had 4 goals against; 1 by C, 0 by E (who scored no goals), ∴ 3 by A. And since A had no goals against, ∴ A vs. D was 3–0. And since A scored 7 goals, A vs. E was 4–0. And since D scored 4 goals, D vs. E was 3–0.

Complete Solution

	A	B	C	D	E
A	×	×	0–0	3–0	4–0
B	×	×	3–1	×	1–0
C	0–0	1–3	×	1–1	×
D	0–3	×	1–1	×	3–0
E	0–4	0–1	×	0–3	×

33. 6 Teams

A diagram will help:

	A	B	C	D	E	F
A	×	×	✓	×	×	✓
B	×	×	✓			✓
C	✓	✓	×	✓	✓	✓
D	×		✓	×		✓
E	×		✓		×	✓
F	✓	✓	✓	✓	✓	×

(i) F played 5 and got 7 points; ∴ either 3 W, 1 T, 1 L or 2 W, 3 T. But it is not possible to win 3 matches and only score 2 goals; ∴ F had 2 W, 3 T. Scores must have been 1–0, 1–0, 0–0, 0–0, 0–0; but we cannot tell yet which were the 2 sides against which F won.

(ii) C lost 1 match and got 7 points. ∴ C must have played everyone (5 matches) with 3 W, 1 T, 1 L. ∴ C and F played everyone, and A, who only played 2 matches, played no one else.

(These facts have been indicated in diagram; the reader is advised to insert other facts as they are discovered in his own diagram.)

(iii) Total of A, B, C, D, F games is 19. Total of all games must be even, since each game occurs twice. ∴ E played an odd number — not 1 as C and F played everyone, and not 5 as E did not play A. ∴ 3. ∴ 11 games were played. B played 4 matches, and since B did not play A, ∴ B played C, D, E, F. ∴ E's games were against B, C, and F; and E did not play D. We know now who played whom.

(iv) 11 games are played and 2 points are awarded for each game, ∴ total of points is 22. Total so far is 20 (3 + 7 + 3 + 7), and since we know that A won 1 game, ∴ A must have gotten 2 points, and E none.

(v) A vs. F must be lost (F lost none), ∴ score 0–1 (see (i)), ∴ A vs. C a win and score 4–1.

(vi) C lost 1, tied 1 and won 3. Lost match was vs. A and drawn match must have been vs. F. (F lost none) and score in C vs. F was 0–0. (See (i).) And C won against B, D and E.

(vii) D played 3 and got 3 points; one match (vs. C) was lost, ∴ D won 1 and drew 1. D cannot have won against F who lost no matches, ∴ D vs. B was won and D vs. F tied. D only scored 1 goal, ∴ D vs. B was 1–0, D vs. F was 0–0, and D vs. C was 0–5.

(viii) E scored no points and ∴ lost their 3 games (against B, C and F). E vs. F must have been 0–1 (see (i)), and F vs. B must have been 0–0 (see (i)).

(ix) B scored only 1 goal, ∴ score in B vs. E (which B won) was 1–0. ∴ score in B's 4th game (against C) was 0–3.

(x) C had 7 goals against, 4 by A, 0 by B, 0 by D, 0 by F, ∴ 3 by E. E had 7 goals against, 1 by B and 1 by F, ∴ 5 by C.

Complete Solution

	A	B	C	D	E	F
A	×	×	4–1	×	×	0–1
B	×	×	0–3	0–1	1–0	0–0
C	1–4	3–0	×	5–0	5–3	0–0
D	×	1–0	0–5	×	×	0–0
E	×	0–1	3–5	×	×	0–1
F	1–0	0–0	0–0	0–0	1–0	×

34. D Was Dumb

1. A played 3 games, ∴ A played everyone. A vs. C must have been a win for A (C lost both games played). And since A won 2 games and scored only 2 goals, ∴ *score in A vs. C was 1–0. And scores in A's other 2 matches were 1–0 and 0–0.*

2. ∴ score in C's other game was 3–5 (see C's totals of goals for and against), ∴ C's other game was not against B, who only scored 4 goals, ∴ *it was against D, and score was 3–5.*

3. Total of games played must be even, since each game appears twice. ∴ D must have played 1 game or 3. But we know that D played C (see above) and A (who played everyone). ∴ *D played 3.*

4. D won against C (see above), and lost against A, who won 2, and against B who tied against A and won the other match.

5. A diagram will help, with results so far.

	A	B	C	D
A	×	T 0–0	W 1–0	W 1–0
B	T 0–0	×		
C	L 0–1		×	L 3–5
D	L 0–1		W 5–3	×

C only played 2 games, ∴ C did not play B. B's other match was against D and score was 4–3 (see B's totals of goals).

Complete Solution

A vs. B	0–0,	B vs. D	4–3,
A vs. C	1–0,	C vs. D	3–5.
A vs. D	1–0,		

35. A Long Leg

1. No one can play more than 2 games. ∴ as figures given are all wrong, A must have played 2 and B must have played 1. Since total of games played must be even (for each game appears twice), ∴ C must have played 1. Since B only played 1 and did *not* win 1 (figure given), ∴ B won *none*. Since C played 1 and did *not* tie 0 (figure given), ∴ C tied 1. ∴ we have as correct figures:

	Played	Won	Lost	Tied	Goals for	Goals against	Points
A	2						
B	1	0					
C	1			1			

2. Clearly A played both B and C, but B and C have not played each other. ∴ C's tied match is against A. We know that B did *not* get 1 point (figure given). ∴ B's match was not tied. And we know that B did not win a game. ∴ B lost against A. ∴ results are:

	A	B	C
A	×	W	T
B	L	×	×
C	T	×	×

3. Consider score in C's tied game against A. It cannot be 0–0, for we know that C did *not* score no goals (figure given). And it cannot be 1–1, for C did *not* have 1 goal against, ∴ *it must be 2–2* (not more than 5 goals in each game).

4. Consider score in B's lost game against A. B did score goals (figure given), ∴ B scored 1 or 2 goals. Since B lost there must have been at least 2 goals against, but these cannot have been 3, for this is the figure given. ∴ B had 2 or 4 goals against.

5. We know that A scored 2 goals against C, and 2 or 4

against B. But we know that A did *not* score 6 goals altogether (figure given). ∴ A must have scored 2 goals against B, and score was therefore *2–1*.

Complete Solution
    A vs. B   2–1,
    A vs. C   2–2.

36. Uncle Bungle Has His Moment

(i) C's match must have been against A and the score must have been *2–2*, A's other game was against B. B lost this one (*1–2*).

(ii) To get the correct table we must keep using the fact that the figures given are wrong.
1. A did not play 2 games, ∴ A played 1. C did not play 1 game, ∴ C played 2. ∴ B played 1 (total of games must be even).
2. C did not win 'none' or lose 'none'; ∴ C won 1, and lost 1. B did not win 'none', ∴ B won against C. ∴ C won against A. We also know that goals given are all wrong.
3. Facts so far can be shown in these 2 tables:

	Played	Won	Lost	Tied	Goals for	Goals against
A	1	0	1	0	not 4	not 3
B	1	1	0	0	not 1	not 2
C	2	1	1	0	not 2	not 2

	A	B	C
A	×	×	L
B	×	×	W
C	W	L	×

4. *Consider B's match against C*
   Possible scores (bearing in mind that B did *not* score 1, and that there were *not* 2 goals against): *2–0, 2–1, 3–0, 3–1*. (Remember that not more than 4 goals were scored in any game.)
5. *Consider A's game against C. Possible scores: 0–1, 0–2, 0–4, 1–2.*
6. Consider now that C's goals for are *not* 2, and that C's goals against are *not* 2. *Possible combinations are:*

B vs. C	2–0	2–1	3–0	3–1
A vs. C	—	1–2	0–1	0–2
			0–4	0–4
				1–2

7. We know that Uncle Bungle is able to give all details from his knowledge of the score in a match in which 2 goals were scored. The only scores with 2 goals are B vs. C (2–0) and A vs. C (0–2). But there is no possible score which goes with B vs. C (2–0). ∴ *it must be A vs. C (0–2) and in this case the score in B vs. C is 3–1.*

Complete Solution
   (i) A vs. B   2–1,
        A vs. C   2–2.
  (ii) A vs. C   0–2,
        B vs. C   3–1.

37. Four Figures False

(i) *Consider A's figures.*

2 W's and 1 T makes 3 played, not 2. ∴ there is at least one mistake in these figures.

Also *no* Goals for, cannot be right with 2 Wins. At least 2 changes must be made for these figures to be consistent with each other.

(ii) *Consider C's figures.*

There must be at least 1 mistake in P, W, L, T. The only way in which these can be put right with one change (remembering that P cannot be more than 2 for "3 teams are each to play each other once") is for T to become 0 instead of 2. But 'Points' would then be 0 and not 1. ∴ at least 2 changes must be made in C's figures.

(iii) We know that there are only 4 mistakes, ∴ there must be 2 in A's figures, 2 in C's figures, *and B's figures must be correct.*

(iv) Consider figures reading downwards.
1. Number of games played must add up to an even number, ∴ there is at least 1 mistake in P.
2. Number of games won must equal number lost; ∴ there is at least 1 mistake in W and L.
3. Number of matches tied must be even (for each match appears twice); ∴ there is at least 1 mistake in T.
4. Total of Goals for, must equal total of Goals against; ∴ there is at least 1 mistake in Goals.

But this makes at least 4 mistakes, ∴ *figures of 'Points' must all be correct.* ∴ total of points = 6.

But this must equal total of P (2 points for each match, each match appears twice in P). ∴ *C's P must be 2 and not 1.*

(v) We know there is *one* mistake in T (but B's figures are correct). ∴ *C's T must be 1 and not 2.*

(vi) *Other 2 mistakes must be A's.*

There must be a change in W or L to make their totals the

same. Since we know that A played 2 and tied 1, this can only be a *change in A's W from 2 to 1*.

(vii) Last mistake must be in A's Goals for or against. At present total for is 3, and total against is 9. To make them equal to each other we must *change A's Goals for from 0 to 6*.

(viii) Correct table is:

	Played	Won	Lost	Tied	Goals for	Goals against	Points
A	2	1	0	1	6	2	3
B	2	1	1	0	3	6	2
C	2	0	1	1	0	1	1

A and C tied 1, obviously against each other. And as C scored no goals, score was 0–0. C lost its other match (against B); and A won its other match (against B). A diagram will help.

	A	B	C
A	×	W	T 0–0
B	L	×	W
C	T 0–0	L	×

(ix) Score in A vs. B must be *6–2* (from goal totals), score in C vs. B must be *0–1* (from goal totals).

Complete Solution

(i) Correct table is:

	Played	Won	Lost	Tied	Goals for	Goals against	Points
A	2	1	0	1	6	2	3
B	2	1	1	0	3	6	2
C	2	0	1	1	0	1	1

(ii) Scores are: A vs. B 6–2,
                 A vs. C 0–0,
                 B vs. C 1–0.

38. Uncle Bungle Predicts

1. Let us find first the detailed results which arise from the first correct table. B vs. D was a tie, and as this was D's only match we can tell that the score was $1-1$.

A table will help

	A	B	C	D
A	×			×
B		×		T $1-1$
C			×	×
D	×	T $1-1$	×	×

This tie was D's only match. ∴ C's 2 games were against A and B. ∴ A, who played 1 game, did not play B.

2. Score in A's game (against C) was $3-1$, ∴ score in C's other game (against B) was $0-2$. We now know the score in each game.

	A	B	C	D
A	×		W $3-1$	
B		×	W $2-0$	T $1-1$
C	L $1-3$	L $0-2$	×	
D		T $1-1$		×

3. Consider now Uncle Bungle's predictions.

By comparing the 2 tables in the question we can deduce predicted score in each game.

B vs. A was 0–3,
C vs. D was 0–1,

and since D vs. C was 1-0, ∴ (looking at D's goals), *D vs. A was 0-0*.

4. We know that in the predictions both the result and the score of each side are wrong. Since 0 appears on at least one side in all of them, ∴ *correct* score can in no case be 0-0; i.e., at least 1 goal is scored in each game. ∴ not more than 3 goals in any game (since total is 5).

5. *Consider D vs. A.* Correct result is not a draw, and each side scored, ∴ score must have been *1-2 or 2-1*.

6. ∴ only 2 goals were scored in other 2 matches.

∴ *B vs. A must have been 1-0,*
and *C vs. D* "   "   " *1-0.*

7. If D vs. A were 1-2, then D's total of 'Goals for' would be 2. But they are given as 2 in Uncle Bungle's predictions

∴ D vs. A cannot be 1-2,

∴ *D vs. A was 2-1*.

Complete Solution

A vs. B	0-1,	B vs. C	2-0,
A vs. C	3-1,	B vs. D	1-1,
A vs. D	1-2,	C vs. D	1-0.

39. Some Ham Humor

The information given in the table is that A did *not* play 3, win 1 etc.

(i) The possibilities for A, reading down, are:

Played	Won	Lost	Tied
2	2	2	2
1	0	0	0

As W, L, T are each 2 or 0, it is not possible for A to have played 1. ∴ the situation for A must be:

	Played	Won	Lost	Tied
	2	2	0	0
or		0	2	0
or		0	0	2

(ii) The possibilities for B, reading down, are:

Played	Won	Lost	Tied
3	3	3	3
1	1	2	2
	0	1	1

None of the alternatives make it possible for B to have played only 1. It is not possible for B to have won 3 because it is not possible for B to have lost 0 and tied 0. Similarly other possibilities can be eliminated. And as B is down as having 3 points we know that B did *not* get 3 points; ∴ B could not have won 1, lost 1 and tied 1. ∴ possibilities are:

	Played	Won	Lost	Tied
	3	0	2	1
or		0	1	2

(iii) The possibilities for C, reading down, are:

Played	Won	Lost	Tied
3	3	3	3
1	2	2	1
	1	1	0

∴ it is not possible for C to have played only 1. And we are left with:

Played	Won	Lost	Tied
3	2	1	0
or	1	2	0
or	1	1	1

If C won 2 and lost 1 they would get 4 points, but we know that C did not get 4 points for all the figures in the table given are incorrect. ∴ position for C is:

Played	Won	Lost	Tied
3	1	2	0
or	1	1	1

(iv) The possibilities for D, reading down, are:

Played	Won	Lost	Tied
2	2	2	1
1	0	1	0

But we now know that A played 2, B 3 and C 3, ∴ number of games played by D must be even (to make total even). ∴ D played 2.

And we also know that D did not get 0 points (figures given), ∴ D did not lose both games. ∴ position for D is:

Played	Won	Lost	Tied
2	0	1	1

(v) It will be useful now to collect the correct figures which we have:

	Played	Won	Lost	Tied	Points
A	2	2	0	0	
		0	2	0	
		0	0	2	
B	3	0	2	1	
		0	1	2	
C	3	1	2	0	
		1	1	1	
D	2	0	1	1	1

If A won 0, greatest total for wins would then be 1, and total for losses would be at least 3, but total of wins must equal total lost. ∴ A won 2 (the only alternative). ∴ A tied 0.

(vi) Suppose B lost 2 and tied 1, then C can not also have tied 1 (this would make total of 3 tied games, but it must be even); ∴ C must have lost 2 and tied 0.

But this makes the total lost 5, and the total won 3, which is impossible. ∴ B did *not* lose 2 and tie 1. ∴ B must have lost 1 and tied 2. ∴ C must have won 1, lost 1 and tied 1 to make the total of tied matches even. ∴ figures are now:

	Played	Won	Lost	Tied	Goals for	Goals against	Points
A	2	2	0	0			4
B	3	0	1	2			2
C	3	1	1	1			3
D	2	0	1	1			1

(vii) We now want to find who played whom and the result of each game. A table will help:

	A	B	C	D
A	×	W	W	×
B	L	×	T	T
C	L	T	×	W
D	×	T	L	×

B tied 2, and C and D tied 1 each; ∴ B vs. C and B vs. D were tied. B's other match (against A) was lost; A's other match (won) must have been against C (who played everyone) and A did not play D. And D's other game against C was lost.

(These results have been inserted in diagram).

(viii) We are told that only 1 goal was scored in a game played by B; B tied 2 games (in which 2 goals or no goals were scored); ∴ score in B's other game (against A) was 0–1.

(ix) Consider A's other game (against C). We know that total of A's goals was not 4 (figure given), ∴ A vs. C was not 3–0. And we know that total against A was not 0 (figure given), ∴ the only possible score for A vs. C is *2–1*.

(x) *Consider B.* B vs. A was 0–1 and other 2 games were tied. If score in both games were 0–0 B's total Goals for would be 0, but we know that it is not 0 (figure given). If score in one game was 0–0 and in the other 1–1, total goals against B would be 2, but we know that is not so. ∴ score in each tied match was 1–1.

(xi) Consider C's other match (against D). We know that C won, ∴ D scored 0 or 1 goal. But total of goals against C is *not* 3 (figure given). ∴ D scored 1 goal against C (we know that A and B between them scored 3 goals against C). ∴ score in C vs. D was 2–1.

Complete Solution

(i) *Correct Table.*

	Played	Won	Lost	Tied	Goals for	Goals against	Points
A	2	2	0	0	3	1	4
B	3	0	1	2	2	3	2
C	3	1	1	1	4	4	3
D	2	0	1	1	2	3	1

(ii) A vs. B   1–0,
    A vs. C   2–1,
    B vs. C   1–1,
    B vs. D   1–1,
    C vs. D   2–1.

40. Uncle Bungle Baffled

1. It is not possible for B to lose a match and have no goals scored against it. ∴ the mistaken figure is either in B's 'Lost' or in B's 'Goals against'. ∴ all other figures are correct.

2. D must have played 1 or 3 to make the total of games played even. But D tied 2, ∴ D played *3* and not 1.

3. A played 3, won none and tied 2, ∴ *A lost 1*. A and D can have tied each other but each of them must also have tied someone else. ∴ C and E must each have tied a match, and therefore neither of them won one. We now have 4 games lost (one each from A, B, C and E) but only 3 possible games won, two from B and one from D. But total of games won must be equal to total lost. ∴ *the mistake in B's line is 'Lost 1.'*

4. This can only be put right by making B's games lost *none*, and B's games won 2. We now have:

	Played	Won	Lost	Tied	Goals for	Goals against
A	3	0	1	2	3	4
B	2	2	0	0	3	0
C	2	0	1	1	1	5
D	3			2		2
E	2	0	1	1	0	

And we see that in order to make total Won equal total Lost, D must have won 1, and *lost none*.

5. We now want to find the result of each game. A diagram will help.

	A	B	C	D	E
A	×	L		T	
B	W	×		×	
C			×		
D	T	×		×	
E					×

A vs. D was obviously a tie, but we cannot yet tell whether A's other tied game was against C or E.

6. Since B won both its games and D lost none, ∴ B did not play D. A's lost game was not against C who won none, nor against E who won none, ∴ it can only have been against B.

(Results discovered so far have been inserted in diagram.)

7. C's lost game is against B or D (no one else won). Since C lost 1 and tied 1, and goals for were 1 and against 5, ∴ C's lost game was either 0–4 or 1–5. But B only scored 3 goals, ∴ C's lost game was against D. D's 3rd game was against E, and was a tie. ∴ A's other drawn match was against C, and A did not play E. C played against A and D, but only played 2 games, ∴ C did not play against B or E. B's other game (won) was against E, and we now know the result of every game.

8. E scored no goals, ∴ score in tied game was 0–0, and in lost game 0–?. B has no goals against in either game (against A and E). As A had 3 goals for and 4 against, and lost one and tied 2, ∴ A's lost match was by a margin of one goal. ∴ A vs. B was 0–1.

9. Since B vs. A was 1–0, ∴ B vs. E was 2–0, C vs. A (tie) is 0–0 or 1–1 (C only scored 1 goal). ∴ A's other tied game (against D) is 3–3 or 2–2. But D only had 2 goals scored against, ∴ *A vs. D was 2–2*. And ∴ *A vs. C was 1–1*. ∴ C's other game (against D) was *0–4*. And we now know the score in every game.

Complete Solution

A vs. B	0–1,	B vs. E	2–0,
A vs. C	1–1,	C vs. D	0–4,
A vs. D	2–2,	D vs. E	0–0.

41. A Muddy Muddle

1. Number of matches played, without B, is 20. ∴ B's *Points would have to be 2 or 4* if these figures are correct. (Total Points must be even for each game appears twice).

But B is down as having 3 L, with 3 points. ∴ *B must have played 5* (to get 3 points it is necessary to win one and tie one). ∴ there must be a mistake in either B across *or* Points down. ∴ *all other figures are correct*.

2. E is down as having played 3, won 2 and got 6 points. But this is not possible. ∴ there is a mistake in E across. ∴ the mistake must be in E's games played. This must be 4 or 5 to produce 6 points.

3. We know that figures given for B are correct. ∴ B must have played 5 (see (1)). ∴ *E must have played 4* to make total of games played even. ∴ correct table now looks like this:

	Played	Won	Lost	Tied	Goals for	Goals against	Points
A	4			0		3	
B	5		3		6	5	3
C	3	2			2	5	
D	5	0		3	0	5	
E	4	2			7	4	6
F	5				4		6

4. B, as we have seen, must have won 1 and tied 1 to get 3 points. Since C won 2 and had 2 goals for and 5 against, ∴ C lost their 3rd match. D lost 2 (5 played — 3 tied). E played 4, won 2 (4 points) and got 6 points altogether; ∴ E must have tied their other 2 matches. D tied 3, ∴ against B, E and F. And F must also have tied against E. ∴ F tied 2, and to get 6 points they must also have won 2, and ∴ lost 1.

5. We have still to fill in A's won and lost. Total won without A is 7, and total lost without A is 7. And since A tied none and total won must equal total of all lost, ∴ A won 2 and lost 2.

(The reader is advised to fill in these details in his own table as in (3).)

6. A table for results and scores will help:

	A	B	C	D	E	F
A	×	✓	×	W✓	✓L	✓
B	✓	×	✓W 5–0	✓T	✓L	✓
C	×	✓L 0–5	×	✓W 1–0	×	✓W 1–0
D	✓L	✓T	✓L 0–1	×	✓T	✓T
E	✓W	✓W	×	✓T	×	✓T
F	✓	✓	✓L 0–1	✓T	✓T	×

We know that B, D and F played everyone. C only played 3, ∴ C did not play A or E. A played 4, ∴ A played E as well as B, D and F. And we now know who played whom.

7. D tied 3 — against B, E and F (as no one else tied a game). ∴ D's other games, against A and C were both lost. As E and F tied 2, they must also have tied each other. E tied 2 and won their other 2 games. ∴ E vs. A and E vs. B were won.

8. C won 2 and lost 1; goals for were 2 and against 5. ∴ C's won games were both 1–0, and lost game was 0–5. We know that C beat D (1–0), and C's lost game was either against B or against F. But F only scored 4 goals altogether, ∴ C's lost match (0–5) was against B, and C vs. F was 1–0.

(Facts discovered so far have been inserted in diagram.)

9. We know that F won 2, and as F lost to C, ∴ F's won games can only have been against A and against B. And A's other win was against B. We now know the result of every game.

10. D scored no goals, ∴ D vs. B, D vs. E and D vs. F were all 0–0. We know that D vs. C was 0–1, and as D had 5 goals against, ∴ D vs. A was 0–4.

11. F scored 4 goals: none against C or D, at least 1 against A and 1 against B (both games won), ∴ not more than 2 against E (E vs. F a tie). E scored 7 goals: not more than 2 against F (see above), ∴ at least 5 against A or B (both won).

A had 3 goals against: at least 1 by F (who won), ∴ not more than 2 by E. ∴ B had *at least 3 goals* scored against it by E.

But B only had 5 goals against it, at least 1 by A and 1 by F (who won), ∴ not more than 3 by E. ∴ *E scored 3 goals against B*.

12. B's other 2 goals against must have been 1 by A and 1 by F, and score in each case was 0–1. ∴ B vs. E was 1–3.

13. A had 3 goals against; none by B, none by D, ∴ 3 by E and F between them.

Let x be number of goals scored by E against A.
Let y " " " " " " F against A.
    Then $x + y = 3$.

Let score in tied game between E and F be p–p. Then considering E's goals for
$x + 3 + 0 + p = 7$, or $x + p = 4$.
And considering F's goals for
$y + 1 + 0 + 0 + p = 4$, or $y + p = 3$.
∴ by subtraction $x - y = 1$.
∴ $x = 2$ and $y = 1$, and $p = 2$.

14. ∴ score in E vs. F was 2–2.
    "   " F vs. A was 1–0.
    "   " E vs. A was 2–1.

Complete Solution

A vs. B	1–0,	B vs. C	5–0,	C vs. D	1–0,
A vs. D	4–0,	B vs. D	0–0,	C vs. F	1–0,
A vs. E	1–2,	B vs. E	1–3,	D vs. E	0–0,
A vs. F	0–1,	B vs. F	0–1,	D vs. F	0–0,
				E vs. F	2–2.

42. No Light of Truth

1. We know that all the figures in the table are wrong. ∴ possible correct figures for A, reading down, are:

Played	Won	Lost	Tied
3	3	3	3
2	2	2	2
1	1	1	1
			0

Possible combinations, reading across, are:

Played	Won	Lost	Tied
3	2	1	0
	1	2	0
	1	1	1
2	1	1	0

But the points given for A are 2, ∴ we know that A did *not* get 2 points, ∴ the possibilities can be reduced to:

Played	Won	Lost	Tied
3	2	1	0
	1	1	1

2. Possible correct figures for B, reading down, are:

Played	Won	Lost	Tied
4	4	4	4
2	2	3	3
1	1	2	2
	0	1	1

Possible combinations, reading across, and remembering that B did *not* get 1 point, are:

Played	Won	Lost	Tied
4	2	1	1
	1	2	1
	1	1	2
	0	2	2
	0	1	3

3. Possible correct figures for C, reading down, are:

Played	Won	Lost	Tied
3	3	3	3
2	2	2	2
1	1	1	1
	0		

Possible combination remembering that C did *not* get 1 point, are:

Played	Won	Lost	Tied
3	1	1	1
	0	1	2

4. Possible correct figures for D, reading down, are:

Played	Won	Lost	Tied
4	4	4	4
2	3	2	3
1	2	1	2
	1	0	1

Possible combinations, remembering that D did *not* get 3 points, are:

Played	Won	Lost	Tied
4	3	0	1
	2	1	1
	2	0	2
	1	1	2
	1	0	3

5. We now know that the number of games played by A, B, C, D must be 3, 4, 3, 4 respectively. ∴ number of games played by E must be *even* (to make total even). It cannot be 4 (figure given), ∴ it must be 2. ∴ possible combinations for E are:

Played	Won	Lost	Tied	Points
2	1	1	0	2
	2	0	0	4

But as E's points are given as 2 they must be 4, and E's figures are:

Played	Won	Lost	Tied	Points
2	2	0	0	4

6. Possibilities now are:

	Played	Won	Lost	Tied
A	3	2	1	0
		1	1	1
B	4	2	1	1
		1	2	1
		1	1	2
		0	2	2
		0	1	3
C	3	1	1	1
		0	1	2
D	4	3	0	1
		2	1	1
		2	0	2
		1	1	2
		1	0	3
E	2	2	0	0

7. D played everyone (4 games). E won both its games. ∴ E beat D. ∴ the three lines which show D as losing none are not possible.

8. We are told that 4 games are tied. This is only possible with the above figures if B tied 3 (against A, C and D) and if C and D also tied each other. The table in (6) now becomes:

	Played	Won	Lost	Tied
A	3	1	1	1
B	4	0	1	3
C	3	0	1	2
D	4	1	1	2
E	2	2	0	0

9. A table showing detailed results will help:

	A	B	C	D	E
A	×	T			×
B	T	×	T	T	L
C		T	×		×
D		T		×	L
E	×	W	×	W	×

B and D played everyone (4 games) and E only played 2, ∴ E played B and D and beat them both. And B tied their other matches.

(Results so far have been inserted in table.)

10. By elimination the game which D won was against A. And the game which C lost was against A. We now know the result of each game.

11. The game in which B played in which one side scored 3 (and ∴ the other 0) must be the game which B lost to E. (B tied its other matches.) E must have had 1 goal scored against them by D (the total given for E's goals against is 0), ∴ E vs. D is 2–1.

12. We know that B scored 2 goals more than C. In B vs. E, B scored no goals, in B vs. C (tie) B and C scored the same number of goals, ∴ B scored 2 more against A and D than C did against A and D. But B can only score 1 against each A and D (both tied), ∴ score in B vs. A and B vs. D was 1–1 in each case; C vs. A was 0–?, and C vs. D was 0–0.

13. Total of goals given *for* B is 3; ∴ B did not score 3 goals, ∴ B vs. C must have been 0–0. Total of goals given against D is 3 ∴ D vs. A cannot be ?–0, ∴ D vs. A is 2–1. Total of goals *against* C is 1, ∴ C vs. A must have been 0–2, *or* 0–3. Total of goals given *for* A is 5, ∴ A vs. C must be 2–0. We now know the scores in every game.

14. Total of goals given against D is 3, ∴ D vs. A cannot be ?–0, ∴ *D vs. A is 2–1*. And we know the score in every game.

Complete Solution

A vs. B	1–1,	B vs. C	0–0,	C vs. D	0–0,
A vs. C	2–0,	B vs. D	1–1,	D vs. E	1–2.
A vs. D	1–2,	B vs. E	0–3,		

43. Zanies, Yodellers and X's

1. XC cannot be right. If C lost 3 games at least 3 goals must have been scored against.

2. YB cannot be right. If B lost a game at least 1 goal must have been scored against.

3. YF cannot be right. Number of games played is less than total of won, lost, tied.

4. ZD cannot be right. Number of games played is less than total of won, lost, tied.

5. ZE cannot be right. If E lost a game at least 1 goal must have been scored against.

6. *Consider Z.* If B won 5 games (i.e., beat all the others), A cannot be right, (at least 1 goal must have been scored against). Similarly F cannot be right, and we already know that D and E are wrong. Also if B won 5 it cannot be correct that C lost none. ∴ if B won 5, the other 5 are all incorrect. But we know that 2 out of the 6 are correct. ∴ *B cannot have won 5, and ZB is incorrect.*

7. Since YB and ZB are both wrong, ∴ *XB is right* (see data.) ∴ B won 0; ∴ it cannot be correct that F lost 5. ∴ *XF is incorrect.* And since YF is also incorrect, ∴ *ZF is right.*

8. YD is not compatible with XB (which we know to be correct), because, if D lost 5, B must have won at least 1. ∴ *YD is incorrect.* And since ZD is also incorrect, ∴ *XD is right.*

9. Since XB and XD are both right, ∴ other 4 in X are all wrong. ∴ XE and ZE are both wrong, ∴ *YE is right.*

10. We now know which is right for B, D, E and F, and we have to discover whether YA and ZC are right or ZA and YC.

The correct figures for B, D, E and F are as follows:

	Played	Won	Lost	Tied	Goals for	Goals against
A						
B	3	0	2	1	0	7
C						
D	4	1	0	3	6	2
E	2	0	2	0	0	6
F	2	1	0	1	5	0

11. If ZA and YC were correct, then total of games played would be $3 + 3 + 3 + 4 + 2 + 2$ (i.e., 17). But since each game appears twice, total of games played must be even. ∴ ZA and YC cannot be correct; ∴ *YA and ZC are correct.* ∴ the correct details of games played, won, lost etc., are as follows:

	Played	Won	Lost	Tied	Goals for	Goals against
A	5	2	2	1	9	9
B	3	0	2	1	0	7
C	2	2	0	0	5	1
D	4	1	0	3	6	2
E	2	0	2	0	0	6
F	2	1	0	1	5	0

12. We now want to find who played whom, and the score in each game. Results can be inserted in a table, thus:

	A	B	C	D	E	F
A	×			T	W –0	
B		×		T	×	
C			×	×	×	
D	T	T	×	×	W –0	T
E	L 0–	×	×	L 0–	×	×
F				T	×	×

13. D tied 3; A, B and F each tied 1. ∴ D vs. A, D vs. B and D vs. F were all tied.

14. D played 1 other game and won it. This could not have been against C who lost none, ∴ against E; and D did not play C.

15. E played 2 (one against D) and lost them both. Other game must have been against A who played everyone; and E scored no goals in either game.

(Facts discovered so far are in diagram; the reader is advised to insert other facts as they are discovered in his own diagram.)

16. A played everyone. We know that B vs. A was not a tie, and as B won none it must have been a win for A. ∴ A's other 2 games (against C and F) were lost. C played 2 games (one against A) and won them both. C's other match could not have been against F who lost none; ∴ against B.

17. Since B only played 3 games, all accounted for, ∴ B did not play F. We now know who played whom and the result of each game.

18. B scored no goals; ∴ B vs. D was 0–0, and B vs. A and B vs. C were both 0–?.

19. C had 1 goal against; not by B, ∴ by A.

20. F had no goals against, ∴ F vs. D (a tie) was 0–0, and F vs. A was 5–0.

21. D had 2 goals against, 0 by B, E or F, ∴ 2 by A and D vs. A was 2–2. ∴ D's score against E was 4–0.

22. ∴ E vs. A was 0–2; A scored 4 goals against B (4–0); A vs. C was 1–2; and B vs. C was 0–3.

Complete Solution

A vs. B	4–0,	B vs. C	0–3,
A vs. C	1–2,	B vs. D	0–0,
A vs. D	2–2,	D vs. E	4–0,
A vs. E	2–0,	D vs. F	0–0.
A vs. F	0–5,		

44. Straight as a Die

(i) Consider 11 across and 8 down.
Since 11 across must be odd, ∴ 8 down is the square of an odd number. But 8 down cannot end in 5 (for 11 across would not be prime), ∴ it must end in 1 or 9.

Since sum of digits is less than 11, if second digit were 9, first one would have to be 1. But we know that 7 down is even, ∴ second figure of 11 across cannot be 9 *and must be 1*.

(ii) First figure of 11 across must be even, and sum of digits is more than 5. ∴ first digits must be 6 or 8. But 81 is not prime, ∴ *11 across is 61*.

(iii) Digits of 6 across must be 1, 2, 3, 4 (all different, sum is 10); but we do not know in what order. ∴ first digit of 7 down cannot be more than 4. But sum of digits of 7 down is 19. ∴ *first digit must be 4, and second 9*.

(iv) ∴ first digit of 6 across cannot be more than 3. But if it were less than 3 it would not be possible for each digit of 1 down to be greater than the preceding one. ∴ *it must be 3*, and 1 down is *1 2 3* –.

(v) First figure of 8 down is 1 or 2. But there is no perfect square 2–1. ∴ 8 down is *1 2 1*. And last figure of 6 across must *be 2*.

(vi) Consider 5 down. Since second digit is 2, 1 2 5 and 7 2 9 are possible. But digits of 9 across are all different, ∴ *it must be 1 2 5*. 10 down could be 2 7 or 6 4. But not 2 7 because all digits of 9 across are different. ∴ *6 4*.

(vii) Last digit of 1 down is odd, and greater than 3. Since digits of 9 across are all different it can only be 7.

(viii) 1 across could be 1 2 2 or 1 8 3. 4 across could be 2 3 or 2 9. But since sum of digits of 2 down is greater than 11, 1 across must be 1 8 3 and 4 across must be 2 9.

(ix) 3 down is the same 2 digits as in 4 across (2 and 9); and as 5 across is odd, ∴ *3 down is 2 9*.

Complete Solution

I	2			3
I	8	3		2
4			5	
2	9		I	9
6	7	8		
3	4	I	2	
9				10
7	9	2	5	6
	11			
	6	I		4

45. Straight As A Cork-Screw

A starting point is provided by 6 across in conjunction with 4 down.

6 across must be the square of an odd number (see 4 down); ∴ it must be 25, 49 or 81. The only 3 figure cubes which end in 5, 9 or 1 are 125 and 729 (the cubes of 5 and 9).

Two diagrams will help, in which we can put these two possibilities. Let us call them A and B.

A.

	1	2	
3			4 1
5			 2
		6 2	5

B.

	1	2	
3			4 7
5			 2
		4	9

Consider 3 across. In A this must be 7531, in B it must be 1357. (The reader is advised to put these and other figures in the diagram as they are discovered.)

3 down must now be 729 in A, and 125 in B.

Consider 1 down. In A sum of digits = 8; and since 2nd digit is 5 and all the digits of 5 across are even, ∴ 1 down must be 152.

In B sum of digits = 18; and since 2nd digit is 3, and all digits of 5 across are even, ∴ 1 down must be 738 or 936.

Consider 5 across. In A this must be 2262.

In B we have 2nd figure as 6 or 8, but if 8, the 3rd figures would have to be 0 to make the sum 12, and we know that there are no 0's. ∴ 2nd figure is 6, and first figure of 1 down is 9. ∴ 5 across is 2622.

2 down   In A the first 2 digits are 43.
              In B the first 2 digits are 45.
We have now filled every square in A and B.

But consider *1 across* (Sum of digits is 13).

In A this is not true, but in B it is. ∴ B gives us the correct solution.

Complete Solution

	1	2	
	9	4	
3			4
1	3	5	7
5			
2	6	2	2
		6	
5		4	9

46. Uncle Bungle's Cross Number

It is obvious that we cannot fill anything in until we have found the incorrect clue. We must look for discrepancies.

(i) (6–9) can only be 27 or 64. But (7–8–9) cannot end in 4 (it is the square of an odd number), and it cannot end in 7 (no square ends in 7). ∴ either (6–9) or (7–8–9) is wrong.

(ii) (2–5–8) is odd. ∴ (8) is an odd digit. But no 3 figure square of an odd number has an odd digit for its second digit. ∴ either (2–5–8) or (7–8–9) is wrong. But as we know that only *one* clue is incorrect it must be (7–8–9). ∴ all the other clues are correct.

(iii) From (i), (6–9) must be 27 or 64. ∴ (6–5–4) must be a square starting with 2 or 6. ∴ it must be 225, 256, 289, 625 or 676.

(iv) Consider (1–4–7) and (2–5–8). Since (2–5–8) is odd, ∴ (1–4–7) is odd. And (2–5–8) must be (1–4–7) multiplied by an odd number — 3 or possibly 5, but no more because of (1–2–3).

(v) (4–5–6) is 522, 652, 982, 526, or 676.

Consider (4) (5), bearing in mind that (2–5–8) is 3 or 5 times (1–4–7). If (4) (5) is *52*, then we must have – 5 – × 3 producing – 2 – (not – 5 – × 5 as this would make (2) at least 7 and (3) would not be less than 10). But – 5 – × 3 cannot produce – 2 –. ∴ *522 and 526 not possible.*

If (4) (5) is 65, then we have – 6 – × 3 producing – 5 – (*not possible*).

If (4) (5) is 67, then we have – 6 – × 3 producing – 7 – (*not possible*).

If (4) (5) is 98, then we have – 9 – × 3 producing – 8 – (*which is possible*).

And since this is the only possibility, ∴ (4–5–6) must be 982. ∴ (6–9) is 27.

(vi) (1) must be 1, and (2–5–8) must be 3 times (1–4–7) — if 5 times (3) would be more than 9. ∴ (2) must be 5, and (3) is 9.

(vii) (2-5-8) is odd and divisible by 3. ∴ 585 is only possibility; ∴ (1-4-7) is 195.

Complete Solution

(7-8-9) is the incorrect clue.

1
   1
	2
	   5
		3
		   9

4
   9
	5
	   8
		6
		   2

7
   5
	8
	   5
		9
		   7

47. A Three-Dimensional Cross Number

It is obvious that no single clue can produce a unique answer. We have to consider clues which are linked together.

(i) (21, 23, 25), (7, 16, 25) and (3, 14, 25) are such clues. From (21, 23, 25) (25) must be even. ∴ (7, 16, 25) is an even cube, and can only be 216 or 512 (the cubes of 6 and 8). But from (3, 14, 25) it cannot be 512. ∴ (7, 16, 25) is *216*.

We cannot yet find the other digits in (21, 23, 25) or (3, 14, 25).

(ii) *Consider (13, 14).* From (3, 14, 25) (14) is less than 6, and greater than 1. ∴ (13, 14) must be 25 or 64.

(iii) *Consider (7, 13, 19).* We know it is $2\frac{2}{6}-$.

If 2nd digit were 6, the 3rd digit would be 10, which is impossible. ∴ 2nd digit is 2, and (7, 13, 19) is *222*. ∴ (13, 14) is *25*.

(iv) *Consider (3, 12, 21).* This is a multiple of 25, ∴ since there are no 0's it ends in 5, and second figure must be 2 or 7. Either 625 or 175 could make sum of digits 13, but from (3, 14, 25) it is not possible for (3) to be 6. ∴ (3, 12, 21) is *175*.

(v) (10, 13, 16) is now -21. ∴ it must be *921*. (10, 11, 12) must be *987*.

(vi) *Consider (4, 13, 22).* Since (13) is 2, ∴ (4) must be 1, and number is *124*.

(vii) *Consider (1, 4, 7).* (1) must be 3, 6, or 9. But from (1, 2, 3), (1) + (2) = 16. But this is not possible if (1) is 3 or 6; ∴ (1) must be 9 and (2) is 7.

(viii) Consider (3, 15, 27) of form $1-x$; and (7, 17, 27) of form $2-x$. From (7, 17, 27) x is odd, ∴ (3, 15, 27) must be 121 or 169. But if x is 9 sum of digits of (7, 17, 27) cannot be 11 (no 0's). ∴ (3, 15, 27) is *121*, and (7, 17, 27) is *281*.

(ix) (16, 17, 18) must be 183, 186 or 189. From (12, 15, 18) it must be 183 or 189. ∴ (9, 18, 27) is $-\frac{3}{9}1$; and (18) must clearly be 3, and (9) must be greater than 3.

(x) *Consider (3, 6, 9) and (7, 8, 9)*. (3, 6, 9) must be an even square. ∴ 144 or 196. But 196 not possible, since sum of digits of (7, 8, 9) is greater by 2 than sum of digits of (3, 6, 9). ∴ (3, 6, 9) must be *144*, and (7, 8, 9) is *254*.

(xi) *Consider (22, 23, 24)*. (23) cannot be 5 (see (5, 14, 23)). ∴ it must be *6*, and (24) is *8*. From (5, 14, 23) (5) must be *8*.

(xii) *Consider (20, 23, 26)*. Sum of digits = 14. But (26) must be at least 7, ∴ (20) is (1) and number is *167*.

Complete Solution

Top

1 9	2 7	3 1
4 1	5 8	6 4
7 2	8 5	9 4

Middle

10 9	11 8	12 7
13 2	14 5	15 2
16 1	17 8	18 3

Bottom

19 2	20 1	21 5
22 4	23 6	24 8
25 6	26 7	27 1

48. A Treble Cross Number

We obviously need 3 different diagrams — let us call them A, B and C.

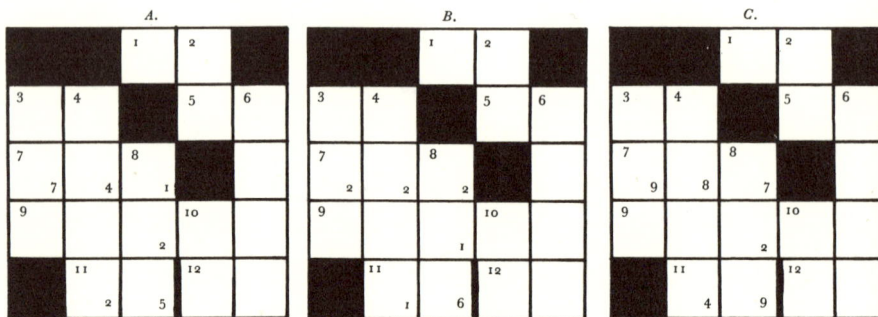

(i) First we need a starting point — some clue for which there are 3 different answers, or some combination of clues to which there are 3 different answers.

This is provided by 8 down (a 3 figure cube) and 11 across (a 2 figure square).

2 figure squares are 16, 25, 36, 49, 64 and 81; 3 figure cubes are 125, 216, 343, 512, 729.

Only three of these cubes end in a figure in which a square can end, namely 125, 216, 729. ∴ we put these figures in squares A, B, C. 11 across in A must be 25; in B it can be 16 or 36; in C it must be 49 (figures inserted).

(ii) Consider 4 down. In A 2nd figure must be 4; in B it must be 2 or 6; in C it must be 8.

Consider 7 across. In A 1st figure must be 7; in B 1st figure must be 2 and 2nd figure 2 (if 2nd figure were 6 1st figure would be 10, which is impossible); in C first figure must be 9. ∴ in B 1st figure of 11 across must be 1 and not 3.

(Further figures should be inserted in diagrams as they are discovered.)

(iii) Consider 3 down. 1st figure of 9 across cannot be greater than 5, ∴ in B 1st figure of 3 down must be 9, and last figure 5. ∴ 3 across must be 96, and 4 down is 6231.

In A 2nd figure of 9 across must be 1, 3 or 4 (not 5, see 4 down), ∴ 2nd figure of 3 across must be 2, 6 or 8. If 2 then 3 across would be 32 and 3rd figure of 3 down would be 6 — but this is not possible (see 9 across). If 2nd figure of 3 across were 6, then 3 across would be 16 or 96. 3 down would then be 178 or 970, neither of which is possible.

∴ 2nd figure of 3 across is 8, and 4 down is 8442, and 3 across is 48, and 3 down is 475.

In C 2nd figure of 9 across must be 1, 3 or 4. ∴ 2nd figure of 3 across must be 2, 6 or 8. If 8 then 3 across would be 48, which is the same as 3 across in A; but we are told that "in no case is the complete answer to a clue the same in different solutions"; ∴ not 8. If 2nd figure of 3 across were 6, then 3 across would be 16 or 96. 3 down would then be 196 or 99–? (impossible to make digits add up to 16). ∴ 2nd figure of 3 across in C is 2, and 1st figure must be 3, and 3 down is 394.

(iv) We can now fill in 12 across as 24 in A, 48 in B and 16 in C.

(v) 6 down in A must be 1234, and in C it must be 3456 (no other possibilities).

In B 5678 and 2468 are not possible (see 9 across); – – 58 is not possible; and the only possibility is *1248*.

(vi) ∴ 9 across in A is 54213; in B it is 53124; and in C it is 41235.

(vii) 5 across in A is 51; in B it is 21; and in C it is 93.

(viii) 2 down in A is 45; in B it is 72; and in C it is 99.

(ix) 1 across in A is 54; in B it is 77; and in C it is 69.

Complete Solution

A.

		1		2		
			5		4	
3		4		5		6
	4		8		5	1
7			8			
	7		4	1		2
9				10		
	5		4	2	1	3
	11			12		
		2	5		2	4

B.

		1		2		
			7		7	
3		4		5		6
	9		6		2	1
7			8			
	2		2	2		2
9				10		
	5		3	1	2	4
	11			12		
		1		6	4	8

C.

		1		2		
			6		9	
3		4		5		6
	3		2		9	3
7			8			
	9		8	7		4
9				10		
	4		1	2	3	5
	11			12		
		4		9	1	6

The magic numbers are: 12, 24 and 31.

49. Top Hats and Tails

It will be convenient to put the data in 'If . . . then' form, thus:
1. If tail coat, then over 16.
2. If top hat, then over 15.
3. If watching baseball on Saturday P.M., then top hat or tail coat or both.
4. If umbrella or over 16 or both, then not sweater (∴ if sweater, then neither umbrella nor over 16).
5. If watching baseball, then sweater.

∴ *if watching baseball*
   from (5) sweater.
     ∴ from (4) not umbrella, not over 16.
     ∴ from (1) not tail coat.
     ∴ from (3) top hat; ∴ from (2) over 15.

Complete Solution
   Between 15 and 16; wearing a sweater and a top hat; not wearing a tail coat; not carrying an umbrella.

50. The Older the Better

1. *Consider B(2)*. If this is true, then B makes a true statement; since C is older than A, C must also make at least one true statement. ∴ A must be the least truthful and makes 2 false statements. If B(2) is false, then A is older than C, and A is the only one who can have made 2 true statements. ∴ *A's statements are either both true or both false.*

2. ∴ A cannot be between B and C in age. ∴ A cannot be younger than B, and older than C. ∴ A(1) and A(2) cannot both be true. ∴ *A(1) and A(2) are both false, and A must be the youngest of the three.* ∴ *B(2) is true.*

3. *Consider B(1) and C(2)*. If these were both true, A's age would be 42. But we know that A is the youngest and that their ages are between 43 and 31 inclusive. ∴ A's age cannot be 42. ∴ B(1) and C(2) not both true. But it follows from the conditions that 3 of the six statements are true and 3 false. We know that A(1) and A(2) and *either* B(1) or C(2) are false, ∴ C(1) must be true.

4. ∴ C is one year older than A. It follows that since A is the least truthful, ∴ C makes one true statement (C(1)) and both of B's statements are true. ∴ B(1) is true, and C(2) is false.

5. ∴ *A must be 36* (we have already seen that A cannot be 42). And since C(1) is true *C is 37*. We know that B is the oldest of the three, but is not 3 years older than A (A(1) false). ∴ *B could be 38, 40, 41, 42 or 43.*

Complete Solution
  A is 36.
  B is 38, 40, 41, 42 or 43.
  C is 37.

51. Orders about Orders

($>$ means 'is greater than'; $<$ means 'is less than.')

(i) From (1): If E is not equal to B, then A = C *and* if A = C, then E is not equal to B.

(ii) From (2): If A $>$ D, then (B $-$ C = A $-$ B).

(iii) From (3): A $>$ C $>$ D.

(iv) From (iii) A $>$ D; ∴ from (ii) (B $-$ C) = (A $-$ B). (These are both positive, since A $>$ C.)

∴ A $>$ B, and B $>$ C.

(v) From (i), if A is not equal to C, then E = B. But we know that A is not equal to C; ∴ E = B.

Complete Solution

1. A
2. = $\begin{cases} B \\ E \end{cases}$
4. C
5. D

52. Who's Sitting Where?

Mr. Binks is not the dentist (who sits in the middle), nor the principal (principal's wife next to him).
∴ Binks is accountant.
Dentist nearer to Dawn than to Bloggs.
∴ Bloggs not dentist. ∴ Bloggs is principal. ∴ Bunn is dentist. ∴ Places from left to right are as follows:

Binks	Mrs.	Bunn	Bloggs
Accountant	Bloggs	Dentist	Principal

Mrs. Bunn must therefore be on Blogg's right (not next to her husband) and Mrs. Binks on Blogg's left.
Dawn is next to the dentist and Rupert has Jane on his left, Elizabeth on his right.
∴ Dawn must be Mrs. Bloggs; Rupert is Bloggs; Jane, on his left, is Mrs. Binks and Elizabeth is Mrs. Bunn.
Dentist is not John, ∴ Binks is John, and Bunn is Ethelred.

Complete Solution

John Binks	Dawn	Ethelred	Jane	Rupert	Elizabeth
Accountant	Bloggs	Bunn	Binks	Bloggs	Bunn
		Dentist		Principal	

53. In-Laws

(It is very important to remember that "every person who is mentioned is one of the five.")

(i) From data B and C are male, A and E female.

(ii) 2 brothers are referred to in (1), (3) and (4). There cannot be 3 different pairs, or even 2 different pairs, as from (i) there are not more than 3 males.

(iii) Suppose there are 3 brothers, B, C, D; then whoever says (1) is of a different, younger generation (A or E, ∴ female); and whoever says (3) is of a different, older generation (E or A, ∴ female). But there is now no one who can be 'my brother's wife' mentioned in (4).

(iv) ∴ there are only 2 brothers and they are B and C. And these two brothers are those referred to in (1) and (3).

(v) We also know that whoever said (3) (call him or her X) is of a generation older than that of B and C, and that whoever said (1) (call him or her Y) is of a generation younger. And there is also the 5th person.

(vi) If (4) is made by someone of the same generation as X then there must be at least 3 people of this generation (the speaker, 'my brother' and 'my brother's wife'). But this would make at least 6 people (these 3, B and C, and one of a younger generation). ∴ 4 cannot be made by someone of the same generation as X. Similarly (4) cannot be made by someone of the same generation as Y. ∴ 4 must be made by B or C and A is the wife of whichever of them did not make it. This is the 5th person referred to in (v).

(vii) From (3), C is the brother whose wife is not mentioned; ∴ A is married to B, and (4) is made by C.

(viii) Since A, B and C are members of the middle generation, ∴ D and E are the members of the older and the younger generation (not necessarily respectively). From (2) E is the member of the older generation, and it must have been E who made statement (3). ∴ E must be the mother of B's wife (i.e., the mother of A). And (2) must have been made by B.

(ix) ∴ D is the member of the younger generation, who made statement (1), and is the son or daughter of C (we cannot tell which).

Complete Solution

Statement (1) made by D.
" (2) " " B.
" (3) " " E.
" (4) " " C.

B and C are brothers.

A is B's wife; E is A's mother. D is C's son or daughter.

54. Only Two True

(i) If (2) and (6) both false C would be seven places higher than D, which is impossible. ∴ either (2) or (6) or both true, ∴ not more than one other remark true.

(ii) If (3) true A 2nd or lower. ∴ (7) true. But (3) and (7) cannot both be true (see (1)). ∴ (3) false. ∴ A higher than B ∴ (5) true.

(iii) Since either (2) or (6) also true, ∴ other remarks (1), (4), (7), must be false, ∴ from (7) A 1st, F 7th.

(iv) ∴ E not 1st and not 2nd or 3rd ((1) is false). D not 7th (F is), ∴ D not 3 places lower than E. ∴ (6) true, ∴ (2) false. ∴ C four places higher than E. ∴ C 2nd, E 6th.

(v) Since (4) false, ∴ B two places lower than G. ∴ G 3rd, B 5th. ∴ D 4th.

Complete Solution

The two true statements are (5) and (6).

The order is:
1. A.
2. C.
3. G.
4. D.
5. B.
6. E.
7. F.

55. Some Buffers Bite

Using obvious abbreviations:
 (i) If not N.C., then since toothless, ∴ B.C. (second condition of membership). But he is not B.C. ∴ *he is N.C.*
 (ii) If N.C., then since under 70, ∴ B.C. (third condition of membership). But he is not B.C. ∴ *he is not N.C.*
 (iii) Since under 70 and B.C., he can be N.C. (third condition of membership). Or since toothless and B.C. he can be '*not* N.C.' (second condition of membership). ∴ *we cannot tell.*

Complete Solution
 (i) Yes. He is a member of the Nashum Club.
 (ii) Yes. He is not a member of the Nashum Club.
 (iii) No. It is not possible to tell.

(A more general, but rather longer, way of solving this problem would be by setting out in a diagram all the sixteen possibilities formed by the combination of B.C., not B.C., over 70, not over 70, etc.; there are 4 classes of which everyone is or is not a member, and therefore 16 possibilities).

56. The Island of Perfection

(i) 5 B. With *Black* stamps the cost will be *4 P.D.*

2 *Green* stamps will bring the weight down to 3 B exactly, and the cost will then be again *4 P.D.* Obviously to send it by red stamps will cost at least *6 P.D.*

(ii) $9\frac{1}{2}$ B. With *Black* stamps the cost will be *8 P.D.*

4 *Green* stamps will bring the weight down to $5\frac{1}{2}$ B, and therefore the cost will be *8 P.D.*

3 *Red* stamps will bring the weight down to $3\frac{1}{2}$ B, and therefore 3 Red stamps are not enough, ∴ cost would be *10 P.D.*

4 *Red* stamps will bring the weight down to $1\frac{1}{2}$ B, and therefore this will be a cheaper way of doing it with red stamps. Cost will be *8 P.D.*

(iii) $10\frac{1}{2}$ B. With *Black* stamps the cost will be *10 P.D.*

4 *Green* stamps will bring the weight down to $6\frac{1}{2}$ B, ∴ cost will be *8 P.D.*

If *Red* stamps are used there is no price between 6 P.D. and 10 P.D.

3 *Red* stamps (costing 6 P.D.) will lessen weight by 6 B's, ∴ weight will become $4\frac{1}{2}$ B's. But at this weight 10 P.D.'s are required. ∴ 6 P.D. not enough, and 10 P.D. more expensive than Green stamps. ∴ least cost will be *8 P.D.*

(iv) $13\frac{1}{2}$ B. With *Black* stamps cost will be *16 P.D.*

5 *Green* stamps will bring weight down to $8\frac{1}{2}$ B, but number of Green stamps then required will be 6. ∴ cost will be *12 P.D.*

Suppose 5 *Red* stamps are used, then weight will be $3\frac{1}{2}$ B and cost will be *10 P.D.*

(v) $15\frac{1}{2}$ B. With *Black* stamps cost will be *20 P.D.*

6 *Green* stamps will bring weight down to $9\frac{1}{2}$ B, and value of stamps required will be 16 P.D. 7 *Green* stamps will bring weight down to $8\frac{1}{2}$ B, and value of stamps required will be 14 P.D. ∴ *14 P.D.* Green stamps are required.

6 *Red* stamps will bring weight down to $3\frac{1}{2}$ B but value of stamps required is then *10 P.D.*

5 *Red* stamps will bring weight down to $5\frac{1}{2}$ B, and value of stamps required is then 10 P.D. ∴ only 5 *Red* stamps are needed. ∴ *10 P.D.* Red stamps are required.

Complete Solution
- (i) *Either* 2 Black stamps
      or 2 Green stamps   Cost 4 P.D.
- (ii) *Either* 4 Black stamps
      or 4 Green stamps   Cost 8 P.D.
      or 4 Red stamps
- (iii) 4 Green stamps   Cost 8 P.D.
- (iv) 5 Red stamps   Cost 10 P.D.
- (v) 5 Red stamps   Cost 10 P.D.

57. Round the Table

It will be convenient to set down and number the separate items of data:

1. Dick and Mr. Smith often play bridge with the Architect's wife and Mrs. Green.
2. The Machine Operator, who is an only child, has Mary on his right.
3. The Politician is sitting nearer to Nancy than he is to Mrs. Brown.
4. Harry is the Architect's brother-in-law, and he has his only sister sitting on his left. The Architect is sisterless.

From (4) Harry is not the Architect, and Harry has a sister. ∴ from (2) Harry is not the Machine Operator (who is an only child). ∴ *Harry is Politician.*

From (3) Politician must be sitting next to Nancy and opposite Mrs. Brown. ∴ *Politician is Brown.* (Each man has a lady on each side of him and the 3rd lady, opposite him, is his wife.)

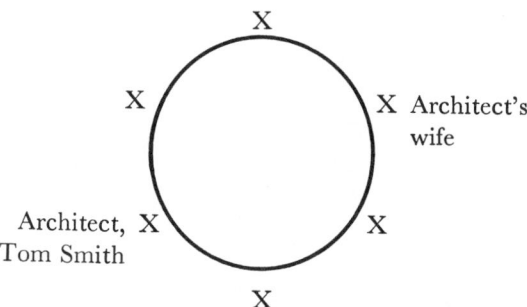

A diagram will help, with Harry Brown, Politician, filled in in any position. From (4) Harry has the Architect's wife on his left. ∴ the Architect is sitting opposite her.

From (1) Architect's wife is not Mrs. Green. ∴ *Architect is*

Smith, and from (1) Dick is not Mr. Smith, ∴ Tom is Smith. (Results so far have been shown on diagram.)

∴ by eliminations *Dick Green is the Machine Operator*, and is sitting on left of Architect's wife (Mrs. Smith).

The Machine Operator has Mary on his right, ∴ *Mary is Mrs. Smith*.

From (3) the Politician has Nancy on his right, ∴ *Nancy is Mrs. Green*. ∴ *Joan is Mrs. Brown*.

Complete Solution

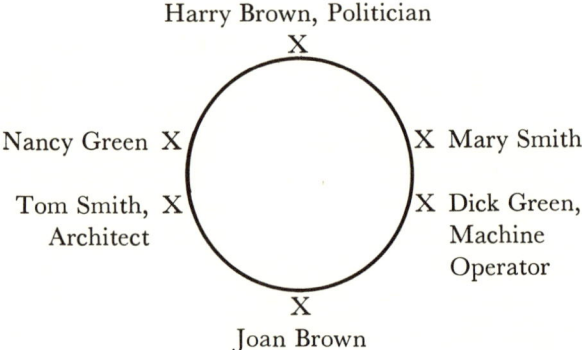

58. The Family Tree of Truth

(i) If Nancy's first statement is false she has at least one brother or sister, ∴ every alternate statement is true, ∴ her second statement is true, ∴ she has a son. But she also has a brother or sister, ∴ all her statements are true. But this is contrary to our hypothesis that her first statement is false. ∴ *her first statement is true*, and as she has no brothers or sisters her second statement is false and her third true. ∴ she must have a child, but it is not John.

(ii) We can argue similarly for Lucy. If her first statement is false (so that she has children), then her second statement must be true. ∴ she has a sister and at least one child, ∴ all her statements are true. But this contradicts our original hypothesis, which must therefore be false. ∴ *Lucy's first statement is true*, her second is false and her third true.

(iii) From Nancy's (true) third remark John has a son. From Lucy's (true) third remark John has a sister (Lucy). ∴ *all John's remarks are true*.

(iv) ∴ John is married to Nancy, James is their son, and Pamela is John's aunt.

(v) ∴ *all James's remarks are false* (Lucy is his aunt, not his sister; Pamela is not his mother; Pamela is John's aunt, not his sister). ∴ he is an only son, and he has no children.

(vi) Pamela's first remark is true (it agrees with John's third). ∴ her third remark is also true. And since Pamela is John's aunt and Lucy is John's sister, ∴ Pamela's second remark is true.

Complete Solution

John: all true.
James: all false.
Nancy: true, false, true.
Lucy: true, false, true.
Pamela: all true.

Pamela is the mother of Nancy and the aunt of John and Lucy, who are brother and sister. John is married to his cousin Nancy, and they have one son, James.

59. The Dowells Do Better

(i) Consider first the statements about how they are related which we know to be true.

From C(1) and B(1) the relationship of B, C and D can be represented thus:

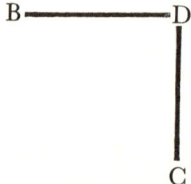

From B(2) we know that the father of B and D was present (making the third generation). And from E(2) we know that there is another member of C's generation.

These two must be A and E, but we do not know which is which.

(ii) Suppose the father of B and D is E, and the brother of C is A.

From D(1) (true) we have:

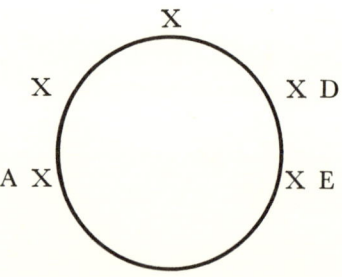

And since E(1) is true, C must be on E's left, and from E(2) A must be on C's left. ∴ we have:

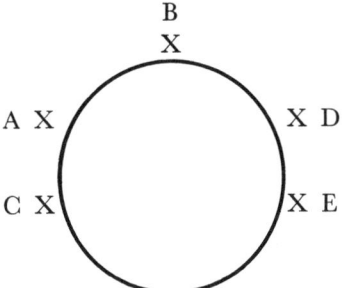

But in this case B(2) *is* not *true* (B is not next to E). ∴ *our assumption is false.*

(iii) Suppose the father of B and D is A, and the brother of C is E. Then since D(1) and A(1) would then both be true, the positions of A, D and E could be represented thus:

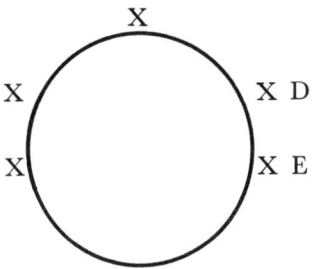

And since (B(2)) B would be sitting next to his father (A), B would be on A's left and C on D's right. All statements are then true, except E(1) and E(2), which is as it should be.

Complete Solution

1. A is the father of B and D, who are brothers, D has two sons, C and E.

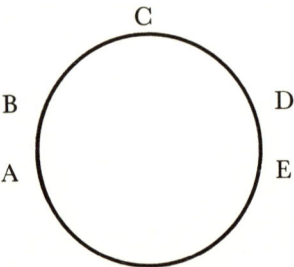

60. Logic about the Logic Prize

Data
S says R won L.
B " J " E.
J " S did not win M.
R " B won F.
Winners of M and L were correct.

(i) Suppose that B is correct (in saying that J won E). Then J is wrong, ∴ S did win M, ∴ S is correct, ∴ R won L, ∴ R is correct, ∴ B won F, ∴ B is wrong.

But this is contrary to our original assumption. ∴ this must have been false, ∴ J did not win E; and B did not win M or L as his speculation has been shown to be wrong.

(ii) Suppose that R is wrong (in saying that B won F). Then B and R are both wrong, ∴ the other two must be right. ∴ R won L (S's speculation). But in this case R would have to be correct, which is contrary to our hypothesis. ∴ R's speculation was correct; ∴ *B won F*.

(iii) Since R is correct, ∴ he did not win E. ∴ by elimination *S won E*. ∴ S was wrong and R did not win L. ∴ by elimination, *J won L*, and *R won M*.

Complete Solution
Smith	won the	English	Prize.
Brown	"	" French	" .
Jones	"	" Logic	" .
Robinson	"	" Mathematics	" .

61. Some Eccentric Sportsmen

Put the data in more compact form, using obvious abbreviations.

If E.S.C., then P in P, and C in C.	(i)
If C in C, then H.C.	(ii)
Not H.C. unless P in P (or if H.C. then P in P).	(iii)
Either E.S.C. or O.A. or both (or if not E.S.C. then O.A., and if not O.A. then E.S.C.).	(iv)
Not both O.A. and C in C (or if O.A., then not C in C).	(v)

(It will be helpful to remember that from 'If x, then y', it is *not* possible to deduce 'If not x, then not y' or 'If y, then x'; but it is possible to deduce 'If not y, then not x'.)

1. *Smith* is not H.C.; ∴ from (ii) *he is not C in C*. From (i), if not (P in P and C in C) then not E.S.C.; but Smith is not C in C, ∴ *he is not E.S.C.* ∴ *from (iv) he is O.A.*

2. *Jones* is C in C. ∴ from (ii) *he is H.C.*, ∴ from (iii) *he is P in P*. From (v) *he is not O.A.;* ∴ from (iv) *he is E.S.C.*

3. *Robinson* is not P in P. ∴ from (iii) *he is not H.C.;* ∴ from (ii) *he is not C in C;* ∴ from (i) *he is not E.S.C.* And ∴ from (iv) *he is O.A.*

Complete Solution
1. Smith has never played Croquet in Czechoslovakia and is not a member of the Eccentric Sportsmen's Club; he is a member of the Oddfellows' Association.
2. Jones is a founder member of the Hoop Club, a member of the Eccentrics' Club, and has played Polo in Patagonia; he is not a member of the Oddfellows' Association.
3. Robinson has not played Croquet in Czechoslovakia, is not a founder member of the Hoop Club or a member of the Eccentric Sportsmen's Club. He is a member of the Oddfellows' Association.

62. Happy Family Unions

A diagram will help, in which results can be entered as they are discovered.

	Lives at	*Engaged to*	*Who lives at*
Mr. Brown			
Mr. Green			
Mr. Black			
Mr. White			

Notice that for each man his name, the name of his house, of his fiancée and of her house are all different, so that if we know three of them we can deduce the fourth.

Consider who lives at the White House. It is not the Browns or the Greens (they are entertained there), nor is it the Whites (data). Therefore it must be the Blacks.

The brother of Mr. Brown's fiancée is not Mr. Green, nor Mr. Black (he was entertained by the Blacks), therefore it is Mr. White. Therefore Mr. Brown is engaged to Miss White.

Mr. Black's fiancée does not live at the Brown House, nor at the Black House (data), nor at the White House (the Blacks live there). Therefore she lives at the Green House. And by elimination, looking at Mr. Black's row, she must be Miss Brown.

Therefore the Browns live at the Green House, and looking at Mr. Brown's row we see that his fiancée, Miss White, must live at the Black House.

Therefore the Whites live at the Black House. The only two ladies left are Miss Black and Miss Green, and Mr. White must therefore be engaged to Miss Green who must live at the Brown House. Mr. Green must be engaged to Miss Black.

Complete Solution

	Lives at	*Engaged to*	*Who lives at*
Mr. Brown	Green House	Miss White	Black House
Mr. Green	Brown House	Miss Black	White House
Mr. Black	White House	Miss Brown	Green House
Mr. White	Black House	Miss Green	Brown House

63. Multiplication and Division

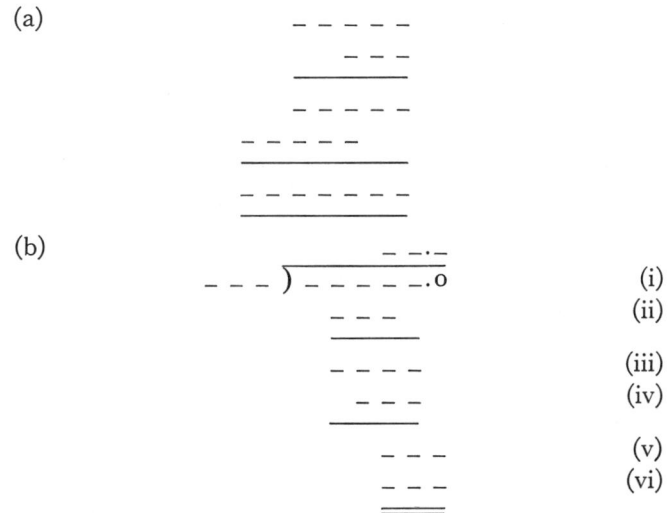

From (a), the middle digit in the smaller number must be 0. (Insert this, and other figures as they are obtained, in a diagram of (a) and (b).)

The last figure in (v) and (vi) must be 0. Suppose divisor ends in 0 and is therefore _ 0 0. (iv) could not then be greater than 900 and (iv) from (iii) would leave at least 100. But it leaves less than 100. ∴ last figure in divisor is not 0. (iii) must be 10 _ _, and (iv) must be 9 _ _ (since subtraction of (iv) from (iii) leaves 2 figures only). (i) must start with 1 and (ii) must start with 9. Since (ii) starts 9 _ and (iii) starts 10, ∴ (i) (by adding up) must start 10.

Divisor must end in an even number or in 5 to go exactly into _ _ 0 ((v), (vi)), and it must be multiplied by 5 or an even number.

∴ divisor is not 9 _ _ .

When divisor (_ 0 _) is multiplied to make 9 _ _ there

cannot be anything to carry from the ten's digit to the hundred's digit, ∴ first digit of divisor goes exactly into 9, ∴ it is 1 or 3.

But we are told there are no 3's, ∴ it is *1*.

Suppose divisor is 105. Then (ii) and (iv) are both 945. (v) must start at least with 6 in order that (iii) shall be 10 _ _, and that (v) shall be divisible by the divisor. If (v) starts with 6 the second figure in (v) would be 3. But there are no 3's. If (v) starts with 8 (the only other possibility with 105 as divisor), there will be found to be a 3 in (a). ∴ divisor is not 105. ∴ (v) and (vi) must be 5 times the divisor. ∴ 5 plus second figure of (iv) must be at least 10, so that there shall be 1 to carry to make 4 figures in (iii). ∴ (iv) must be at least 95 _ . ∴ divisor must be 106 or 108. If 106 there would be a 3 in (v) and in (vi). ∴ divisor is 108.

Complete Solution

```
(a)    10746              (b)           99.5
         108                      108)10746.0
       ─────                            972
       85968                            ────
       10746                            1026
      ──────                             972
      1160568                           ────
      ══════                             54 0
                                         54 0
                                         ════
```

64. 9 Digits Divided by 2 Digits

```
         _ _ _ _ _ _
  _ _ ) _ _ _ _ _ _ _ _ _           (i)
        _ _                        (ii)
        ___
          _ _ _                    (iii)
          _ _                      (iv)
          ___
            _ _ _                  (v)
            _ _                    (vi)
            ___
              _ _                  (vii)
              _ _                  (viii)
              ___
                _ _ _              (ix)
                _ _ _              (x)
                ===
```

(i) must start 10 _ , and (ii) must be 9 _ . (iii) must be 1 _ _ . ∴ (i) must start 100, and (ii) must be 99.

∴ divisor must be 99 or 33 or 11. If divisor were 99, (viii) would be 99, and first figure of (ix) would be 0. ∴ divisor not 99. If divisor were 11, (ix) could not have three figures, for 11 × 9 = 99. ∴ divisor not 11. ∴ *divisor must be 33*.

(iv) and (vi) must both be 99, because there are three figures above, and (v) must be 10 _ . ∴ (iii) must be 109, and (i) starts 10009. Quotient is now 3033 _ 0 _ .

So far we have no 5's (we are told there are 4). (ix) must start with 1 or 2 (33 × 9 = 297).

(viii) must be 33 or 66 and (vii) must be 34, 35, 67, or 68, giving at the most *one* 5. If (vii) were 67 or 68, (v) would be 105, and there would also be a 5 above in (i).

But there would not be enough 5's unless there are also 5's in or resulting from (ix) and (x).

The only 3 figure multiple of 33 which has any 5's is 33 × 5 (165). ∴ (ix) must be 165, and (x) is 165.

Also the last figure in the quotient is 5, and the last figure in (i) is 5. This makes four 5's, ∴ *there must not be any more.*
∴ (vii) cannot be 67 or 68 (see above).
∴ (viii) must be 33 and not 66; and (vii) must be 34.

Complete Solution

```
          3033105
    33)100092465
       99
       ‾‾
        109
         99
         ‾‾
         102
          99
          ‾‾
           34
           33
           ‾‾
            165
            165
            ===
```

65. 7 Digits Divided by 2 Digits

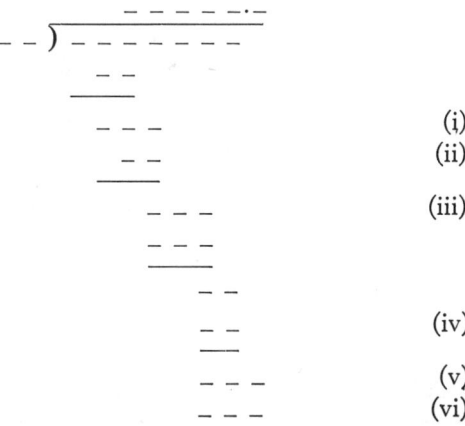

1. Consider (i), (ii) and (iii). 2 digits subtracted from 3 digits leaves 1 digit, ∴ (i) must be 10 _ and (ii) must be 9 _ (but not 90, for 90 from 100 leaves 2 figures).

2. (iv) cannot be 9 _ ; if it were there would be no figure below the 9 after the subtraction. ∴ divisor is not 9 _ , but goes exactly into 9 _ .

3. Since (v) and (vi) end in 0, ∴ divisor goes exactly into _ _ 0. ∴ divisor must be even or end in 5. If divisor were to end in 5 every multiple of it would end in 5 or 0. But we have seen that (ii) cannot be 90, and if (ii) were 95 divisor would not end in 5 (for 95 = 5 × 19). ∴ divisor is even and (ii) must be 92, 94, 96 or 98.

4. If 94 divisor could only be 47, but divisor must be even. ∴ *not 94*.

5. If 98 divisor would have to be 49 or 14; but *not* 49 because divisor must be even, and not 14 because (v) and (vi) would have to be 5 times 14 (to end in 0) and 5 times 14 only has 2 figures. ∴ *not 98*.

6. If 92 divisor 46 (not 23 because divisor must be even); (iii) would then have to start 8 or 9 (92 from 100 or 101), and it would not have been necessary to bring down 3rd figure in (iii). ∴ *not 92.*

7. ∴ (ii) must be *96*. If divisor started with less than 4 (100 − 96) it would not be necessary to bring down 3rd figure in (iii). ∴ *divisor must be 48.*

8. Since last figure of (vi) is 0, ∴ (vi) is 5 × 48, (i.e., 240).

9. (iv) must be 48 or 96. 96 not possible for line above would then have 3 figures, ∴ 48. And line above (iv) is 48 + 24 = 72.

10. Consider (iii). First figure not greater than 4, or it would not have been necessary to bring down third figure. But it must be at least 4 (100 − 96 = 4). ∴ (iii) is 4 _ _ . If 48 into (iii) went 8 times remainder would be more than 7 (48 × 8 = 384). ∴ 48 into (iii) goes 9 times, and (iii) is 439. Rest follows easily.

Complete Solution

```
         22091.5
    48)1060392
       96
       ──
       100
        96
        ──
        439
        432
        ───
         72
         48
         ──
         24 0
         24 0
         ════
```

66. 5 Digits Divided by 3 Digits

Since 3 figures of (i) subtracted from 4 figures above leave 1 figure in (ii), ∴ (i) must be 9 9 _, and line above starts 1 0 0 _.

If (i) was 9 9 0 the subtraction from a 4 figure number would leave 10 or more, ∴ (i) is not 9 9 0.

(iv) must end in 0, ∴ divisor goes exactly into 9 9 _ less than 10 times, and into _ _ 0.

∴ divisor ends in 2, 4, 5, 6, or 8.

To find possibilities divide 9 9 _ by 9, 8, 7, 6, etc.

Divide by 9; no possibilities (remembering that (i) is not 9 9 0).

Divide by 8; 124 is a possibility.

Divide by 7; 142 is a possibility.

Divide by 6; 166 is a possibility.

Divide by 5; no possibility.

If we divide by 4 or less we get a number which is greater than 200, so that when we multiply by 5, as we must do to produce (iv), ending in 0, we get a 4 figure number.

∴ Divisor can only be 124, 142 or 166.

(iv) must be 5 times divisor, ∴ it must be 620, 710, or 830. But last figure of (iii) must be even, because (iii) is a multiple of an even number. And since last figure of (ii) is 0, ∴ second figure of (iv) must be even, ∴ (iv) cannot be 710 or 830, but must be 620.

∴ divisor is *124*.
∴ (i) is 124 × 8 = 992.
∴ 1st figure of (iii) at least 8.

And since (iv) is 620, ∴ (iii) ends in 8 and must be 868 (124 × 7).

Add up from below and we see that (ii) is 930. Add up again and the result is complete.

Complete Solution

```
            80.75
       ―――――――――
    124)10013
        992
        ―――
         93 0
         86 8
         ――――
          6 20
          6 20
          ════
```

67. An Ascending Quotient

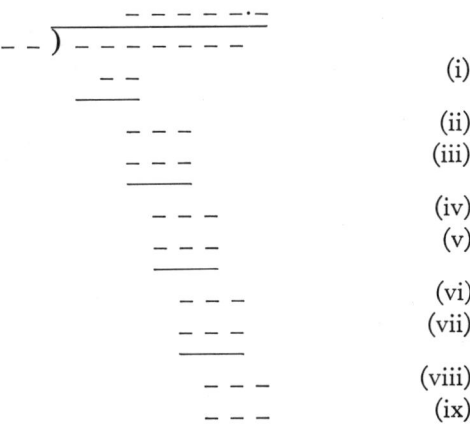

1. (i) (2 figures) is subtracted from 3 figures and leaves 1 figure, ∴ (i) must be between 91 and 99 inclusive. And since divisor goes exactly into (i), ∴ divisor does not end in 0. Figures above (i) must be 10 _ .

2. Last figure in (viii) is not brought down, ∴ it is 0. ∴ divisor goes exactly into _ _ 0.

3. ∴ *either* the divisor is an even number and goes 5 times into (viii), *or* the divisor ends in 5 and goes an even number of times into (viii).

4. *Suppose the divisor is an even number and goes 5 times into (viii).* Then from the conditions of the question it must go respectively once, twice, three times and four times into (i), (iii), (v) and (vii). ∴ divisor must be 92, 94, 96 or 98.

5. Suppose divisor is 92. Then (ii) would be at least 8 _ _ , and it would not be possible for (iii) to be twice the divisor.

239

Similarly since (iii) is twice the divisor it cannot be 94 or 96.
∴ *divisor must be 98.*

6. Situation now is:

```
                  1 0 2 3 4.5
          9 8 ) 1 0 _ _ _ _ _
                  9 8                        (i)
                  ___
                      _ _ _                  (ii)
                      _ _ _                  (iii)
                      _____
                        _ _ _                (iv)
                        _ _ _                (v)
                        _____
                          _ _ _              (vi)
                          _ _ _              (vii)
                          _____
                            _ _ 0            (viii)
                            _ _ 0            (ix)
                            =====
```

We can now fill up (iii), (v), (vii) and (ix) as 98 × 2, 98 × 3, 98 × 4, and 98 × 5. Add up from the bottom *and this solution is complete.*

7. *Suppose now that the divisor ends in 5 and goes an even number of times into (viii).* We know that divisor is _ 5 and that it goes exactly into a number between 91 and 99 inclusive. A consideration of the possibilities (15, 25, etc.) shows that *the divisor can only be 95.*

8. (ii) cannot be less than 5 _ _ . ∴ 3rd figure in quotient must be at least 5. Since (ix) ends in 0, ∴ it must be 95 multiplied by an *even* number. ∴ (iii), (v), (vii) and (ix) can only be 95 × 5, 95 × 6, 95 × 7 and 95 × 8.

9. Situation now is:

```
              1 0 5 6 7.8
        9 5 ) 1 0 — — — — —
              9 5                        (i)
              ───
                    — — —                (ii)
                    — — —                (iii)
                    ───
                    — — —                (iv)
                    — — —                (v)
                    ───
                    — — —                (vi)
                    — — —                (vii)
                    ───
                      — — —              (viii)
                      — — —              (ix)
                      ═══
```

We can now fill up (iii), (v), (vii) and (ix); add up from the bottom *and the solution is complete.*

Complete Solution

(i)
```
                    10234.5
             98)1002981
                98
                ──
                 229
                 196
                 ───
                  338
                  294
                  ───
                   441
                   392
                   ───
                    49 0
                    49 0
                    ════
```

(ii)
$$\begin{array}{r}10567.8\\95\overline{)1003941}\\95\\\hline 539\\475\\\hline 644\\570\\\hline 741\\665\\\hline 76\ 0\\76\ 0\\\hline\hline\end{array}$$

68. An Unconvincing Division

It will be convenient to set a pattern beside the figures given in which the correct figures can be inserted as they are discovered. Thus:

```
           1 9 3                      _ _ _        (i)
     6 6 ) 9 1 2 6      (ii) _ _ ) _ _ _ _        (iii)
           6 8                        _ _          (iv)
           -----                      ---
           1 3 2                      _ _ _        (v)
           1 1 1                      _ _ _        (vi)
           -----                      ---
             2 1 5                    _ _ _        (vii)
             2 1 0                    _ _ _        (viii)
             =====                    ===
```

1. Consider first the subtraction of (iv) from (iii), bearing in mind always, that the figures we are looking for must in every case be *different* from the figures given.

1st figure in (iii) must be 8 or less, 1st figure in (v) must be 2 or more, ∴ 1st figure in (iv) cannot be greater than 6, and it is not equal to 6 (figure given), ∴ it is less than 6.

2. (ii) into (iv) goes at least twice, ∴ (ii) must start with a 2 (not 1 because (ii) would then go into first two figures of (v), which would be absurd, and not more than 2 because first figure in (iv) is 5 or less). And since (v) and (vi) cannot start with 1 (they did before), they must each start with 2. ∴ we have:

```
                 2 _ _              (i)
    (ii) 2 _ ) _ _ _ _              (iii)
                 _ _                (iv)
                 ---
                 2 _ _              (v)
                 2 _ _              (vi)
                 -----
                 _ _ _              (vii)
                 _ _ _              (viii)
                 =====
```

3. Let us try to find second figure in divisor (ii).

(vi) is divisor multiplied by 8 or less (9 is 2nd figure given in (i)). But 8 times 24 or less is less than 2 _ _ . ∴ divisor is 25 or more.

If divisor were 25 every multiple of it would end in 5 or 0. But from figures given in (vii) and (viii) that multiple does not end in 5 or 0. ∴ divisor not 25.

Divisor cannot end in 6 because divisor given (66) ends in 6.

If divisor were 27 (vi) would be 27 times 8 (not 9 because we know 2nd figure in (i) is not 9; and not 7 or less because 27 × 7 is less than 200). But 27 × 8 = 216, and we know that 2nd figure in (vi) is not 1.

If divisor were 29, 2nd figure in (iv) would be 8, but we know it is not. ∴ *by elimination divisor is 28.*

4. (vi) is a multiple of 28, but since it is not 28 × 9 (because 2nd figure in (i) is not 9), ∴ it must be 28 × 8 (224) (28 × 7 is less than 200).

5. ∴ we have:

```
                    2 8 _             (i)
      (ii)  2 8 ) _ _ _ _             (iii)
                    5 6               (iv)

                    2 _ _             (v)
                    2 2 4             (vi)

                    _ _ _             (vii)
                    _ _ _             (viii)
```

We know that (vii) and (viii) do not start with 2 (they do in original). ∴ they start with 1.

∴ they are 28 × 4 (112)
        or 28 × 5 (140)
        or 28 × 6 (168)
        or 28 × 7 (196).

But (viii) cannot be 112 (2nd figure given in (viii) is 1), and (viii) cannot be 140 (3rd figure given in (viii) is 0), and (viii) cannot be 196 (last figure given in (iii) is 6). ∴ (viii) must be *168*.

Complete Solution

$$\begin{array}{r} 286 \\ 28\overline{)8008} \\ 56 \\ \hline 240 \\ 224 \\ \hline 168 \\ 168 \\ \hline \end{array}$$

69. 4 Digits Divided by 2 Digits (all incorrect)

It will be convenient to set a pattern beside the figures given, in which the correct figures can be inserted as they are discovered. Thus:

```
           1 7 9                      _ _ _          (i)
     7 9 ) 9 3 9 8    (ii) _ _ ) _ _ _ _            (iii)
           6 6                          _ _          (iv)
           -----                       -----
           1 8 9                      _ _ _          (v)
           1 5 6                      _ _ _          (vi)
           -----                       -----
             1 3 5                    _ _ _          (vii)
             1 3 0                    _ _ _          (viii)
             =====                    =====
```

1. Divisor goes more than *once* into (iv) (first figure in (i) is given as 1, ∴ it is not 1). ∴ divisor (ii) is less than 50.

2. First figure in (iii) is not 9 (figure given), ∴ it is 8 or less. First figure in (iv) cannot be 8 or 9 (if it were subtraction would not be possible or there would not be a first figure in (v)).

If first figure in (iv) were 7, first figure in (v) could only be 1. But we know it is not 1 (figure given). ∴ first figure in (iv) is not 7, nor 6 (figure given). ∴ *it must be 5 or less.*

3. ∴ divisor starts with 2 or 1.

But if divisor started with 1 it would not be necessary to bring down a figure in (v), for we know that (v) starts with 2 or more. ∴ *divisor starts with 2.*

∴ (v), (vi), (vii) and (viii) all start with 2 (not 1 because that is figure given, not more than 2 for in that case it would not be necessary to have a 3rd figure).

4. Since (iv) starts with 5 or less and we know that first figure in (i) is not 1, ∴ first figure in (i) must be 2.

5. The greatest possible value for the divisor is 29. 29 × 6

is only 174. ∴ second figure in (i) is at least 7 (to bring (vi) up to 2 _ _). But it is not 7 (figure given), ∴ it is 8 or 9.

6. Divisor cannot be 29, for 9 is figure given.

Since second figure in (iv) (which is twice the divisor) is given as 6, ∴ divisor is not 23 or 28. (It cannot be less than 23, for 22 × 9 = 198.) If divisor were 25 every multiple of divisor would end in 0 or 5. ∴ from (vii) and (viii) divisor cannot be 25.

If divisor were 24, (vi) and (vii) would both be 24 × 9 (216) for 24 × 8 is less than 200. But we know that (vi) does not end in 6 (figure given). ∴ divisor is 26 or 27.

7. If divisor were 26, (vii) and (viii) would be either 208 (26 × 8) or 234 (26 × 9). But (vii) cannot be 234 for it is given as 135. And (vii) cannot be 208, for 8 is given as last figure in (iii), which must also be last figure in (vii) and (viii).

∴ divisor is *not* 26.

∴ divisor must be *27*.

∴ (vi) and (viii) must be 216 (27 × 8) or 243 (27 × 9). But (vi) cannot be 216, for 6 is 3rd figure given, ∴ (vi) is 243. And (vii) cannot be 243, for 9 is given as last figure in (i).
∴ (vii) and (viii) are 216. We know that (iv) is 54 (27 × 2). And we now add up from the bottom.

Complete Solution

$$\begin{array}{r} 298 \\ 27\overline{)8046} \\ 54 \\ \hline 264 \\ 243 \\ \hline 216 \\ 216 \\ \hline \end{array}$$

70. 6 Digits Divided by 2 Digits (all incorrect)

It will be convenient to set a pattern beside the figures given, in which the correct figures can be inserted as they are discovered. Thus:

```
              9 2 1 3                  _ _ _ _        (i)
      4 3 ) 2 3 8 9 9 5         _ _ ) _ _ _ _ _ _     (ii)
            1 0 2                      _ _ _          (iii)

              6 9                        _ _          (iv)
              6 4                        _ _          (v)

                5 9                      _ _          (vi)
                2 8                      _ _          (vii)

                3 1 5                    _ _ _        (viii)
                2 1 0                    _ _ _        (ix)
```

1. 3rd figure in (i) must be at least 2. ∴ divisor goes at least twice into (vii) (2 figures). ∴ divisor is less than 50. Divisor does not start with 4 (figure given). ∴ divisor starts with 3 or less.

2. (iii) must start with at least 2, ∴ (ii) must start with at least 3 — it cannot start with 2 (figure given) and the 1st figure of (ii) cannot be less than the 1st figure of (iii). ∴ divisor starts with 3 or more. ∴ *divisor starts with 3.*

3. (ii) must start with 3, and (iii) must start with 2 or 3, and must be at least 291, since (iii) subtracted from three figures leaves only one figure. (iii) is divisor multiplied by 8 or less (since 9 is 1st figure in answer). $36 \times 8 = 288$, ∴ *divisor must be more than 36.* And since $39 \times 7 = 273$, ∴ 1st figure in (i) must be *8.*

4. Since divisor multiplied by 3 produces three figures, ∴ (v) must be divisor $\times$ 1, and (vii) must be divisor $\times$ 2 (to be

different from figure given). ∴ from (vii) divisor is *not 39* (for we know that divisor × 2 does not end in 8). If divisor is 38, then (iii) would be 38 × 8 (304), (not 38 × 9, because 9 is given as first figure in answer; and not 38 × 7 or less, because this would be too small). But we know that 2nd figure in (iii) is *not* 0, because this is the figure given. ∴ divisor is not 38, ∴ *can only be 37*.

5. Situation now is:

```
              8 1 2 _              (i)
      3 7 ) _ _ _ _ _ _            (ii)
            2 9 6                  (iii)
            _____
              _ _                  (iv)
              3 7                  (v)
              ___
                _ _                (vi)
                7 4                (vii)
                ___
                _ _ _              (viii)
                _ _ _              (ix)
                =====
```

37 × 8 = 296, ∴ 1st figure in (i) is 8, and (iii) *is 296* (for reasons already given this is the only possibility).

6. (viii) and (ix) must be the same, since division sums come out exactly. They cannot start with 2 or 3 (figures given), ∴ they must start with 1. ∴ (ix) must be 37 × 3 (111); or 37 × 4 (148); or 37 × 5 (185). But we know that (viii) does not end in 5 (figure given), and we know that 2nd figure in (ix) is not 1 (figure given). ∴ (viii) and (ix) must be 148 (37 × 4). Add up from the bottom and the rest follows easily.

Complete Solution

$$\require{enclose}\begin{array}{r}8124\\37\enclose{longdiv}{300588}\\296\\\hline 45\\37\\\hline 88\\74\\\hline 148\\148\\\hline\end{array}$$

71. A Mistake in the Mistakes

It will be convenient to place a pattern beside the figures given in which the correct figures can be inserted as they are discovered. Thus:

```
           2 3 4 3                    _ _ _ _         (i)
      18 ) 8 1 0 1 2      (ii) _ _ ) _ _ _ _ _      (iii)
           6 9                        _ _            (iv)
           ─────                      ───
           2 2 0                      _ _ _           (v)
           8 0                        _ _            (vi)
           ─────                      ───
             2 0 9                    _ _ _          (vii)
             8 2                      _ _           (viii)
             ─────                    ───
               7 2                    _ _            (ix)
               7 2                    _ _             (x)
               ═══                    ═══
```

1. Consider (vii), (viii) and (ix). 2 figures subtracted from 3 figures leaves 1 figure. ∴ (viii) must be 9 _ and (vii) must be 10 _ . ∴ 2nd figure given in (vii) is correct (2 *o* 9). ∴ *all the other figures given are incorrect.*

2. If (iv) were 9 _ , there could not be a figure below the 9 in (v). ∴ (iv) cannot be 9 _ , ∴ divisor is not 9 _ , but goes exactly into 9 _ . ∴ divisor is less than 50.

3. From figures given for (ii) (18), divisor does not start with 1. If divisor started with 2 it would go 4 times into (viii) (9 _). But 4 is figure given in (i). ∴ divisor does not start with 2.

4. (vi) (2 figures) subtracted from (v) (3 figures) produces 2 figures. ∴ (v) must be 1 _ _ , and (vi) must be the same as (viii). (∴ it is 9 _ .)

If divisor started with 3 it would go 3 times into 9 _ . But since 2nd figure in quotient is given as 3 it does not. ∴ divisor does not start with 3.

5. ∴ divisor starts with 4.

And since it goes exactly into 9 _ , ∴ it must be at least 45. If it were 45 (vi) would be 90, but (vi) is given as *80*, ∴ 2nd figure is not 0. ∴ divisor is not 45.

If divisor were 46, 2nd figure of (viii) would be 2, but it is given as *2*. ∴ divisor is not 46.

(iv) (69) must be divisor, ∴ divisor is not 49.

Divisor is given as *18*, ∴ it does *not* end in 8.

∴ *by elimination divisor must be* 47.

6. ∴ (iv) is 47; (vi) and (viii) are both 94. Since (viii) is 94, ∴ (ix) must be at least 6 _ . ∴ (ix) and (x) cannot be 47, ∴ they must be 94.

Complete Solution

```
        1222
47)57434
     47
     ---
     104
      94
      ---
      103
       94
       ---
        94
        94
        ===
```

72. Addition and Subtraction

1. Let us call the first number (of 5 digits) (a), and the number of 4 digits (b). From the fact that in (i), (a) and (b) produce a number of six digits, the first digit in (a) must be 9 and the result of the addition must start 10.

(Remember that when two rows of numbers are added together there can never be more than 1 to carry.)

Since first figure in result of subtraction is not 9 (figure given), ∴ it must be 8.

2. If first figure in (b) were 6, (b) would be 6 7 8 9. ∴ first figure cannot be more than 6. But it is not 6 for that is figure given. ∴ it is 5 or less. If second figure in (a) were 4, (a) would be 9 4 3 2 1. ∴ second figure cannot be less than 4. But it is not 4, because that is figure given. ∴ it is 5 or more.

3. In subtracting first figure of (b) from second figure of (a) it must be necessary to borrow 1, since, as we have seen, the result of the subtraction starts with 8. ∴ first figure of (b) is greater than or equal to second figure of (a). ∴ considering argument of (2) they must both be 5, and figure below them in subtraction must be 9.

4. Situation is now:

(i)   9 5 _ _ _   (a)
         5 _ _ _   (b)
      ─────────
      1 0 _ _ _ _

(ii) 9 5 _ _ _   (a)
         5 _ _ _   (b)
      ─────────
         8 9 _ _ _

5. Last figure in (a) can be 2 or 1. Last figure in (b) can be 8 or 9. But it is *not* 9 (figure given), ∴ *it is 8*.

And since result of subtracting last figure in (b) from last figure in (a) is *not* 3 (figure given), ∴ last figure in (a) must be 2. ∴ (a) can only be 9 5 4 3 2 and (b) can only be 5 6 7 8.

Complete Solution

The numbers are: 95432
        5678
        ―――

73. Unable, Unhappy and Unwanted

Call them A, H and W:
 (i) If H true, then A(1) false (∴ W not a Sh-Sh) and since A is W-W, ∴ W a Pukka. ∴ W true (H a W-W), ∴ H false. But this is contrary to our assumption; ∴ assumption wrong, ∴ *H false*.
 (ii) If A(2) false (∴ A is not a Pukka), then W must be Pukka (no one else is), ∴ W true, and H is W-W, and A Sh-Sh. ∴ A(1) true. But we have just said that W must be Pukka not Sh-Sh. ∴ our assumption must be wrong. ∴ A(2) true. ∴ A is Pukka, and A(1) is true, ∴ W is Sh-Sh (and W is true). ∴ H is W-W.

Complete Solution
 Unable is a Pukka.
 Unhappy is a Wotta-Woppa.
 Unwanted is a Shilli-Shalla.

74. The Silent C

(i) Suppose A(1) true, then A(3) true (A's remarks must be alternately true and false). ∴ C has said that he is a W-W. But C cannot say this truthfully for W-W's never tell the truth. ∴ C makes a false statement, and is not a W-W. But he cannot be a Sh-Sh (A is), nor a Pukka (a false statement). ∴ assumption that A(1) is true leads to an impossible situation. ∴ *A(1) false.*

(ii) ∴ A not a Sh-Sh, and since he has made a false statement, not a Pukka. ∴ *A is W-W* and all his statements false.

(iii) If B(1) false, then B is not Sh-Sh, and he cannot be Pukka (false statement), nor W-W (A is). ∴ B(1) cannot be false. ∴ *B(1) is true*, ∴ B is Sh-Sh, and C must be a Pukka.

Complete Solution
  A is a Wotta-Woppa.
  B is a Shilli-Shalla.
  C is a Pukka.

75. The Charming Chieftainesses

Put data in an abbreviated form:
A. 1. D is a Sh-Sh.
2. D is more charming than F.
D. 1. F is a W-W.
2. A is a Pukka.
F. 1. The most truthful of us is the least charming.
2. A is not the most charming.

If D(2) true, then A(1) true, then D(1) false, then F not a W-W. But F must be a W-W if A a Pukka and D a Sh-Sh. ∴ original assumption false, ∴ D(2) false, ∴ A not a Pukka, and D not a Pukka (she has made a false remark). ∴ *F is Pukka.*

∴ D(1) false, ∴ *D a W-W* (2 false).

∴ A a Sh-Sh (by elimination), and since we know that A(1) is false, ∴ A(2) is true. And F(1) and F(2) are both true. From A(2) and F(2) D is 1st for charm. From F(1), F is 3rd.

Complete Solution

Attractive is Shilli-Shalla (1st false, 2nd true).
Delectable is Wotta-Woppa.
Fascinating is Pukka.

	Charm:	
	1st	Delectable,
	2nd	Attractive,
	3rd	Fascinating.

76. Ringing Is Not All Roses

(i) A says his number is 468; B says A's number is 587. Since there is a difference of 2 between the 2nd figures (6 and 8), both of them must be wrong and *the correct 2nd figure must be 7*. (We are told that any digit which is wrong is *1* more or less than the correct number.)

(ii) It follows from the above that neither A nor B is a Pukka, ∴ *C is a Pukka*, ∴ *C's number is 304*. ∴ figure given in A(2) (403) has 2nd digit correct and other 2 wrong. ∴ *A is a Sh-Sh*.

And since 1st figure of A(2) is wrong, ∴ 1st and 3rd figures of A(1) are correct. ∴ *A's number is 478*.

(iii) *Consider B(1)*. We know that all the digits of 942 are incorrect by 1.

The 1st digit can only be 8, the 2nd digit can be 5 or 3, and the 3rd digit can be 3 or 1. ∴ B's number is either 853, or 851, or 833, or 831.

Complete Solution

A is a Shilli-Shalla; his telephone number is 478.

B is a Wotta-Woppa; his telephone number is 853, or 851, or 833, or 831.

C is a Pukka; his telephone number is 304.

77. Logic Lane

(i) If D(1) is true, then A(2) is true, and E(3) is true. But this is not possible, as one of them is a W-W who never makes a true statement. ∴ *D(1) is false;* and at least one of A(2) and E(3) is false (otherwise D(1) would be true).

(ii) ∴ *D(3) is false* (D must be W-W or Sh-Sh). ∴ E's number is odd. ∴ *E(2) is false.* ∴ *A must be Pukka* (the only one who has not made a false statement), *and all A's statements are true.*

(iii) ∴ A(2) is true, ∴ E(3) is false (see above). ∴ *E is W-W* (two consecutive false statements), ∴ D is Sh-Sh, *and D(1) is false, and D(2) true.*

(iv) Since D(2) is true, ∴ *D's number is 37.* And since A(1) and A(2) are true, and D(1) is false, ∴ A's number is either 40 or 44 (with 40 perhaps not possible because of E(1) (false)).

(v) *Consider A(3).* We know that E's number is odd (see (ii)). ∴ E's number cannot be 37 plus or minus 13.

We know that A's number is 40 or 44. ∴ E's number cannot be A's plus 13. ∴ it is A's minus 13, ∴ it is not divisible by 10, ∴ *A's number must be 40,* and *E's number must be 27.*

Complete Solution

Add is a Pukka; the number of his house is 40.

Divide is a Shilli-Shalla; the number of his house is 37.

Even is a Wotta-Woppa; the number of his house is 27.

78. The Top Teams on the Imperfect Island

(i) C(3) is true — B does say that (B(1)). ∴ C a Pukka or a Sh-Sh and C(1) true. ∴ A(2) false; ∴ B(1) false (A is not a Pukka). ∴ *C must be a Pukka* (no one else is).

(ii) Since C(2) true, ∴ A scored 3 goals against C. Since B(1) false, ∴ B(3) false, ∴ B vs. C not a tie.

(iii) *Consider B(2)*. If true then B a Sh-Sh, but from C(1) (true) the Pukka beat the Sh-Sh's, ∴ B(2) false. And since this leads to a contradiction, ∴ our assumption is false, and *B(2) is not true*. ∴ all B's statements are false and *B is a W-W;* ∴ *A is a Sh-Sh*, and A(1) and A(3) are true.

(iv) Since C(1) is true, ∴ C beat A. From A(1) (true), score in C vs. A must have been 4–3, and C scored no goals against B. ∴ C must have lost to B (C scored no goals and we know that it was not a tie).

(v) Since A(3) is true and B vs. C was not a tie, ∴ B vs. C was 1–0, and from A(3) and C(4) A vs. B was 0–0.

Complete Solution

A is a Shilli-Shalla.
B is a Wotta-Woppa.
C is a Pukka.

A vs. B	0–0,
A vs. C	3–4,
B vs. C	1–0.

79. Some Tribal Wedlocks

(i) If A(2) true, then C's remarks all true or all false. If A(1) true, C(2) false. ∴ if A(1) and A(2) both true, C(1) and C(2) both false. But if A(1) and A(2) true, then C(1) true. ∴ A(1) and A(2) not both true, ∴ A is not a Pukka, ∴ C(1) false, ∴ C not a Pukka. ∴ *B a Pukka* and B(1) and B(2) true.

(ii) From B(1) (true) and C(3) (false) we know that C is not married to D or F. ∴ *C is married to E*.

(iii) If A(2) true, then C not a Sh-Sh, ∴ C(2) false, ∴ C a W-W, ∴ A a Sh-Sh and A(1) false. ∴ A not married to F, ∴ B is married to F (no one else can be). ∴ C(2) true, but this is contradictory, ∴ our assumption is false, ∴ A(2) is false.

(iv) ∴ E (C's wife) is not a Sh-Sh, and E not a Pukka (B(2)), ∴ E is a W-W, ∴ C is not a W-W, ∴ C is a Sh-Sh.

(v) ∴ by elimination A is a W-W and A(1) is false. ∴ B is married to F, and A is married to D. F is not a Pukka (married to one) and not a W-W (E is), ∴ *F is a Sh-Sh*. ∴ *D is a Pukka*.

Complete Solution

Arthur (Wotta-Woppa) is married to Dulcy (Pukka).

Bartholomew (Pukka) is married to Fanny (Shilli-Shalla).

Clarence (Shilli-Shalla) is married to Ermintrude (Wotta-Woppa).

80. The Immortal Wolla

(i) If B(1) true then A a Pukka, B a Sh-Sh (he makes a true remark) ∴ C a W-W. Since A is a Pukka, ∴ A(3) true, ∴ C's currency is the Wolla. But C says this (C(3)), and we know that C(3) must be false. ∴ our original assumption must be false, and *B(1) is not true.* (∴ *B(3) is also false.*)

(ii) ∴ A is not Pukka, and B is not Pukka (B(1) false). ∴ *C must be Pukka*, and all C's statements true. ∴ *C's currency is Wolla*.

(iii) We know that A's and B's currencies are Blanks and Mounds, and that A and B are W-W's and Sh-Sh's, but we do not know which is which. From B(3) (false), the W-W's currency is worth more than the Sh-Sh's. From C(1) (true), A's currency is worth more than B's. ∴ *A is W-W* and *B is Sh-Sh*.

(iv) Since A is W-W, ∴ A(2) false, ∴ A's currency is the Mound, and B's is the Blank.

(v) From B(2) (true), 3 Mounds = 4 Blanks. From C(2) true, 1 Blank = 3 Wollas. ∴ 3 Mounds = 4 Blanks = 12 Wollas.

Complete Solution

A belongs to the Wotta-Woppas who use Mounds;
B " " " Shilla-Shallas " " Blanks;
C " " " Pukkas " " Wollas.
The rates of exchange are:
3 Mounds = 4 Blanks = 12 Wollas.

81. Imperfect Wives and Daughters

(i) Suppose B(1) is true, then A is a Pukka, *B is a Sh-Sh* (because B makes a true remark), ∴ C must be a W-W. ∴ C(2) is false. ∴ B is not a Sh-Sh. But this is contrary to our supposition. ∴ *B(1) is not true.*

(ii) ∴ A is not a Pukka, and B is not a Pukka (B(1) false), ∴ C is the Pukka. ∴ C(2) is true, and B is a Sh-Sh (with 1st and 3rd statements false, and 2nd true). ∴ A is a W-W. We now know whether each statement is true or false.

(iii) *A diagram will help:*

		Wife			*Daughter*		
		P	W-W	Sh-Sh	P	W-W	Sh-Sh
W-W	A						
Sh-Sh	B	×	×	✓	×	✓	×
P	C						

We know that B(2) is true, ∴ *B's wife is a Sh-Sh*. ∴ B's daughter is *not* a Sh-Sh. A(1) is false, ∴ B's daughter is *not* a P. ∴ by elimination *B's daughter is a W-W*.

(Each fact discovered so far has been inserted in diagram.)

(iv) Since A(2) is false, ∴ *C's daughter is a P*. ∴ C's wife is *not* a P. Since C(3) is true, ∴ C's wife is not a W-W. ∴ *C's wife is a Sh-Sh*.

(v) Since A(3) is false, ∴ A's wife is *not* a Sh-Sh. Since C(1) is true, ∴ A *does* say that his daughter is a Sh-Sh, but A's statements are false, ∴ A's daughter is *not* a Sh-Sh.

(vi) *Consider B(3) (false).* This tells us that A's wife and daughter belong to the same tribe. We know that neither of them is a Sh-Sh; and they cannot both be W-W's, because A is. ∴ *they are both P.*

Complete Solution

Arthur is a Wotta-Woppa; his wife is a Pukka; his daughter is a Pukka.

Brian is a Shilli-Shalla; his wife is a Shilli-Shalla; his daughter is a Wotta-Woppa.

Clarence is a Pukka; his wife is a Shilli-Shalla; his daughter is a Pukka.

82. Competitions in Imperfection

(i) We must first discover to which tribes they belong.
Consider $B(3)$. If this is true, then $C(2)$ true (for C is a Pukka), ∴ A is a W-W. ∴ B is a Sh-Sh (one of the three to each tribe). ∴ $B(1)$ true (since $B(3)$ true), ∴ B is *not* a Sh-Sh. But this is a contradiction. ∴ our original assumption is false, and $B(3)$ *is false.* ∴ $B(1)$ *is false.*

(ii) Since $B(3)$ is false, ∴ C is not a Pukka, ∴ *A is a Pukka* (no-one else is). And $B(1)$ is false, ∴ *B is a Sh-Sh.* ∴ *C is a W-W.* We now know which remarks are true and which are false.

(iii) Since $B(2)$ is true, ∴ B is higher than C in the socially unacceptable (S.U.) competition. Since $C(3)$ is false, ∴ A is more stupid (S) than C. Since $C(1)$ is false, ∴ A is not 3rd in S.U.

(iv) From $A(1)$ B higher for S than for S.U. But we know that B is not 3rd for S.U. (B higher than C). ∴ *B is 1st for S.* ∴ *A is 2nd and C 3rd for S.* And B *not* 1st for S.U.

(v) Since B higher than C in S.U. (from (iii)), ∴ *B 2nd, C 3rd, and A 1st in S.U.*

(vi) From $A(3)$ *A 1st* in P. And from $A(2)$ *C 2nd* and ∴ *B 3rd* in P.

Complete Solution

A is a Pukka.
B is a Shilli-Shalla (false, true, false).
C is a Wotta-Woppa.

Stupidity	*Plainness*	*Social unacceptability*
1. B	1. A	1. A
2. A	2. C	2. B
3. C	3. B	3. C

83. Uncle Bungle on the Island

1. First let us trace U.B.'s thinking.

He thinks that B's wife is a Pukka. ∴ he must think that C who says that B's wife is a W-W is lying.

U.B. thinks (correctly) that the 3 wives belong to different tribes, ∴ he must think that B's statement is false (it is not possible for another wife to be a Pukka).

∴ *he must think that A's statement is true, and that C's wife is a W-W.*

∴ by elimination he thinks *that A's wife is a Sh-Sh.*

2. U.B. thinks that *A is a Pukka.*

He knows that a marriage between W-W's is not allowed, ∴ he thinks that C is not a W-W, ∴ he must think that C is a Sh-Sh, and that B is a W-W.

3. We know that U.B.'s statement is untrue, and that he is incorrect about the tribes to which all six belong.

A diagram will help.

	P	Sh-Sh	W-W	P	*Wives* Sh-Sh	W-W
A	×				×	
B			×	×		
C		×				×

The above indicates that *all* U.B.'s results are wrong.

4. We know that either B or C is a Pukka.

Suppose B.

Then A must be Sh-Sh (C cannot be), and by elimination C must be W-W. From B's statement (true), A's wife is a Pukka; by elimination C's wife is a Sh-Sh, and B's wife is a W-W.

Thus:

	P	Sh-Sh	W-W	P	*Wives* Sh-Sh	W-W
A	×	✓	×	✓	×	×
B	✓	×	×	×	×	✓
C	×	×	✓	×	✓	×

In this case A is the Sh-Sh and A's statement (C's wife is a W-W) is false. But we are told that Sh-Sh's statement is true. (Also the W-W's statement is true instead of false.) ∴ *B cannot be Pukka*.

5. *Suppose C is Pukka.*

Then B's wife is a W-W (C's statement). And we can complete diagram thus:

	P	Sh-Sh	W-W	P	Wives Sh-Sh	W-W
A	×	×	✓	✓	×	×
B	×	✓	×	×	×	✓
C	✓	×	×	×	✓	×

In this case B is the Sh-Sh and B's statement ("A's wife is a Pukka") is true. (Also the W-W's statement is false.) ∴ *C must be Pukka*, and above diagram is correct.

Complete Solution

1. *U.B. thinks that:*

A is a Pukka, and is married to a Shilli-Shalla.

B is a Wotta-Woppa, and is married to a Pukka.

C is a Shilli-Shalla, and is married to a Wotta-Woppa.

2. *In fact:*

A is a Wotta-Woppa, and is married to a Pukka.

B is a Shilli-Shalla, and is married to a Wotta-Woppa.

C is a Pukka, and is married to a Shilli-Shalla.

84. The Tribal Trousers

(i) *Consider B(3).* If true, then E(3) is true, ∴ D(2) is true, ∴ F(1) is true, ∴ B(2) is false (Sh-Sh's statements alternately true and false). ∴ D is a W-W. ∴ D(2) false. But this is a contradiction, ∴ original assumption must be false. ∴ *B(3) false*, ∴ *B(1) false* (B must be either a Sh-Sh or a W-W).

(ii) If E(3) true, then D(2) true, then F(2) true. But if F(2) true, then E(3) false. ∴ *E(3) false*, ∴ *E(1) false*.

(iii) From E(1) (false) *F is a Pukka.* ∴ *B is a Sh-Sh.* (F(1) and D(2) *true*.)

(iv) From F(3) (true) D not a Pukka (husband and wife not the same tribe). ∴ *D a Sh-Sh.* ∴ from D(1) *C is a Pukka.*

(v) From F(2) (true) *E(2) false.* From F(3) (true) *D is married to F.* From C(1) and C(2) (both true) and B(1) (false) E is 10 years older than C, who is older than B who is 6 years older than F. And since the total span of their ages cannot be more than 17 (46 − 29) and their ages are all different, *E must be 46, C 36, B 35, and F 29.*

(vi) ∴ from C(3) *C is married to E.* ∴ by elimination *A is married to B.* ∴ *A(2) true.* And as A is married to B, A cannot be Sh-Sh, ∴ *A is a Pukka.*

(vii) From A(1) (true) *A must be 42* (not 35, the only other multiple of 7 between 46 and 29, because B is 35). From A(3) (true) *D must be 31.*

Complete Solution

	Tribes	Ages	
A	Pukkas	42	
B	Shilli-Shallas	35	A is married to B
C	Pukkas	36	C " " " E
D	Shilli-Shallas	31	D " " " F
E	Wotta-Woppas	46	
F	Pukkas	29	

85. Some Imperfect Wages and Taxes

We must clearly find out first which statements are true and which false.

(i) *Suppose B(1) false.* Then B(3) false (statements alternately true and false); ∴ B is T, and C is not R. ∴ A is R, B is T, C is S. ∴ A(2) false, and C(3) false. ∴ B and C both start with false remarks. But we are told that two of them start with a true remark. ∴ our original assumption must be false. ∴ *B(1) is true* (∴ *C is R*).

(ii) ∴ B(3) is true. ∴ B is not T. ∴ *B is S*, ∴ *A is T*. ∴ A(2) is true, and C(3) is true. ∴ C(1) is true, and other remarks are all false.

(iii) C(1) is true, ∴ B's income after tax is 321. Income after tax is either 9/10 or 3/4 or 2/3 of income before tax. 321 is divisible by 3, but not by 9 or 2. ∴ 321 is 3/4 of B's income, which is therefore *428*. And B's income, since tax is at 25%, is between the incomes of A and C.

(iv) From A(3) (false) and C(2) (false), C pays 10% of his income in tax, and A pays $33\frac{1}{3}$% of his income in tax. From A(1) (false), A's income before tax is either 372 or 376. But A's income must be divisible by 3. ∴ *A's income is 372*.

(v) C's tax is 10% of his income, which is therefore divisible by 10 and larger than B's (428); but no-one's income is more than 450. C's income is 430, 440 or 450. From B(2) (false) it must be *430*.

Complete Solution

A is Tight; his wages are 372 Wollas; he pays tax at $33\frac{1}{3}$%.
B is Scrooge; his wages are 428 Wollas; he pays tax at 25%.
C is Rockerfeller; his wages are 430 Wollas; he pays tax at 10%.

86. The Years Pass

1. *Suppose B(3) true.*

Then C(3) true, and A(1) false. ∴ B and D belong to the same tribe. But this could not be Pukkas as C is a Pukka (we are assuming that B(3) is true) and it is not possible for more than 2 to belong to the same tribe. ∴ C(1) false; but C(1) should be true if B(3) is true. ∴ our assumption is incorrect and *B(3) is not true.* ∴ *B is not a Pukka*, and *C is not a Pukka*. And since B(3) is false, ∴ B(1) is false, and *B is a W-W*. (Sh-Sh's start with a true statement.)

2. A(3) and D(2) cannot both be true (but one of them must be as either A or D is a Pukka). Suppose A(3) is false, ∴ A(1) is false and B and D belong to the same tribe. But if A's statements are false D must be a Pukka and cannot belong to same tribe as B. ∴ A(1) and A(3) cannot be false, ∴ B is 34, and D(2) is false. ∴ *A is Pukka* (no-one else can be).

3. Since D is not a Pukka and A is, ∴ C(1) and C(3) are false. ∴ *C is a W-W* (Sh-Sh's start with a true statement). ∴ *D is a Sh-Sh* (no one else is), and D(1) and D(3) are true.

4. From D(3) A is 1 year older than C, and from A(2) (true) C is older than D. ∴ A is older than C and D. But from B(1) (false) A is not the oldest, ∴ A is not older than B who is 34. ∴ A is 33 or less, and C is 32 or less. D is 11 years younger than C (A(2)), and is over 12, ∴ C's age is a prime (D(1)) *over 23 and less than 33*. It cannot be 29 for this would make D's age 18 and we know from B(2) (false) that D's age is *not* a multiple of 9. The only other possible prime is *31*. ∴ C is 31, ∴ D is 20 (A(2)), A is 32 (D(3)), and B is 34 (A(3)).

Complete Solution
    Ages   A is 32.
              B is 34.
              C is 31.
              D is 20.

Tribes A is a Pukka.
B is a Wotta-Woppa.
C is a Wotta-Woppa.
D is a Shilli-Shalla.

87. A Mixed Marriage

(i) *Suppose C(2) true;* then A is father and B mother, and since C must be Sh-Sh, ∴ C(4) true. B(1) is false, and since B is Ex-Pukka, ∴ rest of B's statements must be true. But if B(4) is true it contradicts our original assumption, which must therefore be false. ∴ *C(2) is false.* ∴ *A(3) is false.*

(ii) If A(2) false then A must be ex-W-W (two consecutive false statements cannot be made by ex-Pukka or by Sh-Sh). ∴ B(1) would be false, and B(4) would be false (there is only one ex-W-W). But B would not then be ex-Pukka (2 false statements) nor Sh-Sh (1st and 4th statements false). ∴ our assumption is impossible. ∴ *A(2) is true.* ∴ A is ex-Pukka, ∴ *A(1) and A(4) are also true.*

(iii) From A(4) (true) B's number is not the greatest; from A(1) (true) Cecil's number is the greatest. ∴ B cannot be Cecil, ∴ *B(2) is false.* And since we know that B(1) is also false (A is ex-Pukka, and must therefore be the mother or the father). ∴ B cannot be Sh-Sh. *C is Sh-Sh* and *B is ex-W-W.* ∴ B(4) is false, ∴ B(3) is true (B makes one true statement); and C(1) and C(3) are both true. And since C(2) is false, ∴ A must be C's mother, not father.

(iv) From A(1) and C(1) (both true), the order of numbers is Cecil, Evelyn, Sidney, with a gap of 10 between the last two. The difference between A's and B's numbers is 22 (A(4)), ∴ A and B cannot be Evelyn and Sidney, but can only be Cecil and Evelyn or Cecil and Sidney. ∴ *Cecil is A.* And C's number is either 10 greater or 10 less than B's, and therefore either 32 less than A's or 12 less than A's. From B(3) and C(3) it can be seen that the only possibilities are for *A to be 66,* and *C 54.* ∴ *B's number is 44,* and *B is Sidney and C is Evelyn.*

Complete Solution

A is the mother, and is an ex-Pukka.
B is the father, and is an ex-Wotta-Woppa.

C is the son, and is a Shilli-Shalla.
A is Cecil; tribal number 66.
B is Sidney; tribal number 44.
C is Evelyn; tribal number 54.

88. Uncle Bungle Fails Again

1. If $D(3)$ *true*, then A(2) true and all B's statements are false. ∴ B(2) false and C is not U.B. And since A is P, and B is W-W, ∴ D must be U.B. And since D(3) is true D(1) must be false, since we know that U.B.'s statements do not agree with any of the tribal rules. But if D is U.B., A is P, and B is W-W, then D(1) is true. ∴ our assumption leads us to a contradiction. ∴ our assumption is false. ∴ $D(3)$ *is false*.

2. If $D(1)$ *true*, then D would be U.B. and C Sh-Sh: (D(1)). ∴ either A or B could be P. But we know that A is not (D(3) false). ∴ B would be P. ∴ B(2) would be true. But B(2) contradicts D(1). ∴ $D(1)$ *is false* and D is not U.B. And since D(1) is false, ∴ $C(3)$ *is false*. ∴ B *must be P*. (C and D have made false statements and A is not P because D(3) is false.) ∴ $B(2)$ *is true* and *C is U.B.* $A(2)$ *is false* (B is P), and $A(3)$ *is false* (U.B. has no wife). ∴ *A is W-W* and *D is Sh-Sh*.

3. The following diagram may help.

	Tribe	Married to	Tribe of wife
A	W-W		
B	P		
C	U.B.		
D	Sh-Sh		

Since C(3) is false and C is U.B., ∴ C(1) must be true (otherwise C's statements would have the pattern of a W-W or a Sh-Sh). ∴ F is not a W-W. Since B(3) is true, ∴ *D is married to F*. ∴ F is not a Sh-Sh, ∴ *F must be a P*.

4. We know that D(2) is true, ∴ E and G same tribe. E and G are wives of A and B (not necessarily respectively), ∴ their tribe cannot be W-W or P. ∴ *their tribe must be Sh-Sh*. And since A(1) is false, ∴ B not married to G. ∴ B must be married to E. ∴ A is married to G.

Complete Solution

B is a Pukka.
D is a Shilli-Shalla.
A is a Wotta-Woppa.
C is Uncle Bungle.
Ellie   is a Shilli-Shalla and is married to B.
Florinda " " Pukka   "   "   "   " D.
Gertie   " " Shilli-Shalla   "   "   "   " A.

89. World Cup for the Wotta-Woppas?

Consider first the games Pl., W, L, T.

(i) B cannot have played 1 game; the total of W, L, T is bound to be more than 1. ∴ B played 3 and the only possibility for W, L, T, is 1, 1, 1.

	Played	Won	Lost	Tied
∴ B	3	1	1	1

(ii) C cannot have played 4 games (they are each to play against each other once), ∴ C played 2. T must be 1, and L at least 1, ∴ W must be 0.

	Played	Won	Lost	Tied
∴ C	2	0	1	1

(iii) D cannot have played 0 games (they lost at least 1). ∴ D played 2. T must be 1, and L at least 1, ∴ W must be 0.

	Played	Won	Lost	Tied
∴ D	2	0	1	1

(iv) *Consider A.* Total of games played must be even (as each game occurs twice), ∴ A played 1 or 3. Total W must equal total L; so far we have 1 W and 3 L; and total T must be even (so far we have 3 T). ∴ we must have

	Played	Won	Lost	Tied
A	3	2	0	1

(v) *Consider Goals.* Total for must equal Total against, as each goal appears twice.

C's and D's 'Goals for' must each be 1 (not otherwise possible to be 1 out). This now makes

$$\text{'Goals for'} = 6 + 4 + 1 + 1 = 12$$
$$\text{and 'Goals against'} = 2 + 1 + 1 + 2 = 6.$$

There are 6 figures which still have to be changed (6 and 4 on one side; 2, 1, 1 and 2 on the other) and it is clear that we can only make the totals equal by subtracting 1 from 6 and 4, and adding 1 to 2, 1, 1 and 2.

(vi) ∴ the *correct* table is:

	Played	Won	Lost	Tied	Goals for	Goals against
A	3	2	0	1	5	3
B	3	1	1	1	3	2
C	2	0	1	1	1	2
D	2	0	1	1	1	3

A table will help.

	A	B	C	D
A	×			T
B		×	T	
C		T	×	×
D	T		×	×

It is easy to see that C did not play D, but that otherwise they all played each other. (Indicate in diagram, as shown.)

(vii) We now want to find the result of each game. Each of the games which A won was by a margin of 1 goal (2 W, 1 T, 5 Goals for, and 3 against). The game which D lost was by a margin of 2 goals (1 L, 1 T, 1 for, 3 against). ∴ A did not beat D, ∴ A vs. D was a tie. ∴ the other tied game was B vs. C.

(Indicate in diagram as shown; the reader is advised to insert other facts, as they are discovered, in his own diagram.)

(viii) A won 2 and therefore beat B and C. B won 1, and therefore beat D. We now know the result of each game.

(ix) In A vs. D score not more than 1–1 (D only scored 1 goal). In A vs. B score not more than 2–1 (B only had 2 goals against). In A vs. C score not more than 2–1 (C only had 2 goals against).

And if scores in any of these games were less than figures above, A's total of 5 goals for and 3 against could not be reached. ∴ A vs. B was 2–1, A vs. C was 2–1, A vs. D was 1–1.

(x) ∴ C's other game, against B, was 0–0, and D's other game, against B, was 0–2.

Complete Solution

All the sides played against each other except that C did not play D.

Results: A vs. B   2–1,   B vs. C   0–0,
A vs. C   2–1,   B vs. D   2–0.
A vs. D   1–1,

90. Some Assorted Tribal Soccer

(i) Figures for A cannot be right; with 3 tied, goals for should equal goals against. Figures for B not right — points wrong. Figures for D not right — points wrong. But figures for C can be right, ∴ *C must be Pukka*, and all figures right.

(ii) *Consider A.* If W-W, true figures would be:

Played	Won	Lost	Tied	Goals for	Goals against	Points
2	1	1	2	6} 4}	5} 3}	4} 2}

But this is not possible, for $W + L + T > Pl$. ∴ A not W-W.
If A Sh-Sh with 1st statement false, true figures would be:

Played	Won	Lost	Tied	Goals for	Goals against	Points
2	0	1	3	6} 4}	4	4} 2}

Not possible — $(W + L + T > Pl)$.

If A Sh-Sh with 1st statement true, true figures would be:

Played	Won	Lost	Tied	Goals for	Goals against	Points
3	1	0	2	5	5} 3}	3

This would have to be 1 W and 2 T, but this would yield 4 points, not 3. ∴ A not Sh-Sh. ∴ *A must be assorted team.*

(iii) *Consider B.* If W-W, true figures would be:

Played	Won	Lost	Tied	Goals for	Goals against	Points
2	1	3} 1}	2} 0}	1	6} 4}	3} 1}

279

This can only be 1 W, 1 L and points would then be 2, not 3 or 1. ∴ B not W-W. If B Sh-Sh with 1st statement true, true figures would be:

Played	Won	Lost	Tied	Goals for	Goals against	Points
3	1	2	$\left.\begin{array}{c}2\\0\end{array}\right\}$	0	$\left.\begin{array}{c}6\\4\end{array}\right\}$	2

But it is not possible to win a match with no goals for.

If B Sh-Sh with 1st statement false, true figures would be:

Played	Won	Lost	Tied	Goals for	Goals against	Points
2	0	$\left.\begin{array}{c}3\\1\end{array}\right\}$	1	1	5	$\left.\begin{array}{c}3\\1\end{array}\right\}$

With 1 L and 1 Point, this is possible.

∴ *B a Sh-Sh*, and the figures must be:

Played	Won	Lost	Tied	Goals for	Goals against	Points
2	0	1	1	1	5	1

(iv) *Consider D.* If D Sh-Sh with 1st statement false, true figures would be:

Played	Won	Lost	Tied	Goals for	Goals against	Points
$\left.\begin{array}{c}3\\1\end{array}\right\}$	0	$\left.\begin{array}{c}3\\1\end{array}\right\}$	0	$\left.\begin{array}{c}4\\2\end{array}\right\}$	2	$\left.\begin{array}{c}3\\1\end{array}\right\}$

If Pl. 1 that would be L and there would be no points, not 3 or 1. If Pl. 3 they would all be L and again no points.

If D Sh-Sh with 1st statement true, true figures would be:

Played	Won	Lost	Tied	Goals for	Goals against	Points
2	1	2	1	3	$\left.\begin{array}{c}3\\1\end{array}\right\}$	2

And $W + L + T > Pl.$ ∴ D not Sh-Sh.

If D W-W, true figures would be:

Played	Won	Lost	Tied	Goals for	Goals against	Points
3 } 1	1	3 } 1	1	4 } 2	3 } 1	3 } 1

This is possible; thus:

Played	Won	Lost	Tied	Goals for	Goals against	Points
3	1	1	1	4 } 2	3 } 1	3

∴ these must be D's figures, but we cannot yet determine goals.

(v) True tables now appear thus:

	Played	Won	Lost	Tied	Goals for	Goals against	Points
A							
B	2	0	1	1	1	5	1
C	2	1	1	0	4	4	2
D	3	1	1	1	4 } 2	3 } 1	3

Bearing in mind that A's figures can only differ by one if at all from those given, we see that A must have played 3 (total Pl. must be even), Won 1 and Lost 0 (to make total W = total L), Tied 2 (to make total Pl. = 3), and scored 4 points (total Pl. is 10, ∴ total points = 10). Table now reads:

	Played	Won	Lost	Tied	Goals for	Goals against	Points
A	3	1	0	2	4, 5 or 6	3, 4 or 5	4
B	2	0	1	1	1	5	1
C	2	1	1	0	4	4	2
D	3	1	1	1	4 } 2	3 } 1	3

(vi) A table will help:

	A	B	C	D
A	×	D		D
B	D	×	×	
C		×	×	
D	D			×

A tied 2 and since B and D each tied 1, A vs. B and A vs. D are ties (mark in diagram as shown).

A and D played 3, and B and C 2, ∴ B did not play C (mark in diagram — it is suggested that other results should be inserted in diagram as they are discovered).

Since A won 1, ∴ A vs. C is W. Since B lost 1, ∴ B vs. D is L. Since C won 1, ∴ C vs. D is W. We now know the result of each match.

(vii) B tied 1 (against A) and lost against D, and since B's goals were 1 for and 5 against, D must have scored *at least* 4 goals against B. But D's goals for are 4 or 2, ∴ D vs. B was 4–0, and B vs. A was 1–1. ∴ D scored no goals in either of their other matches. ∴ D vs. A was 0–0, and D vs. C was 0–?.

C had no goals scored against them by D, ∴ C had 4 goals against them by A.

A's goals against are 3, 4, or 5; 1 by B, 0 by D and less than 4 by C; ∴ it must be 2 or 3 by C. Since C scored 2 or 3 goals against A, ∴ C scored 2 or 1 goals against D (C scored 4 goals). ∴ D vs. C was 0–2 or 0–1. But D's total goals against were 3 or 1, ∴ D vs. C was 0–1. ∴ C vs. A was 3–4. And we now know the score in each game.

Complete Solution

The correct figures are:

	Played	Won	Lost	Tied	Goals for	Goals against	Points
A	3	1	0	2	5	4	4
B	2	0	1	1	1	5	1
C	2	1	1	0	4	4	2
D	3	1	1	1	4	1	3

The scores are: A vs. B 1–1, B vs. D 0–4,
 A vs. C 4–3, C vs. D 1–0.
 A vs. D 0–0,

91. Our Factory on the Island

A diagram (with obvious abbreviations) will help.

	Pu.	W-W	Sh-Sh	P-Sh	W-Sh	P	Q	R	S	T
A					×				×	
B	×	×	×	×	✓				×	
C					×	×	×	×	✓	×
D	×				×				×	
E					×				×	
P										
Q										
R										
S										
T										

1. Since in the long run every married couple makes as nearly as possible the same number of true and false remarks, ∴ Pu's must be married to W-W's, P-Sh's must be married to W-Sh's, and Sh-Sh's must be married to Sh-Sh's.

2. If B(i) true, then D(ii) true, then A(iii) false, ∴ E is a Pu, ∴ E(ii) true, ∴ C(iii) true, ∴ B(i) false. But this is contrary to our hypothesis, ∴ B(i) false, ∴ D is not a Pukka, and B is not a Pukka (he has made a false remark). (These facts have been marked in diagram.)

3. *Consider B(ii)*. If true B must be a W-W (Pu's are married to W-W's), ∴ it cannot be true, for all remarks would be false. ∴ B's wife not a Pukka. ∴ B not a W-W. But B has made two consecutive false statements, ∴ B a W-Sh, and B(iii) must be true. ∴ C is married to S.

(These facts have been marked in diagram. The reader is advised to insert other facts as they are discovered.)

4. Since B is a W-Sh, ∴ C(iii) false, ∴ C not a Pu. ∴ E(ii) false, ∴ E not a Pu. ∴ A must be a Pu. (since no one else is).

5. E(i) is false (we know B is a W-Sh), ∴ E not a Sh-Sh or P-Sh (two consecutive false remarks), ∴ E is a W-W (nothing else he can be). ∴ E(iii) false, ∴ D not married to R.

6. Since A is a Pu., ∴ D(ii) is false.

7. Since S is married to C, and C is a Sh-Sh or P-Sh, ∴ S is a Sh-Sh or W-Sh.

8. A(ii) is true (A is a Pu.), ∴ P not a W-W. ∴ P not married to a Pu. ∴ P not married to A.

9. Since C(iii) is false, ∴ C(ii) must be true (C is either a Sh-Sh or a P-Sh), ∴ R is one of the Shallas. ∴ R married to one of the Shallas. ∴ R not married to A or E.

10. ∴ by elimination R is married to B. We know that B is a W-Sh. ∴ R is a P-Sh.

11. Since D(ii) is false, ∴ D(i) is true (D is either a Sh-Sh or a P-Sh). ∴ Q not a W-W. ∴ by elimination T is a W-W. ∴ T is married to a Pu., ∴ T married to A.

12. ∴ C(i) is true, ∴ C is a P-Sh (his remarks are true, true, false). ∴ by elimination D is a Sh-Sh.

13. We know that S is married to C and that C is a P-Sh, ∴ S is a W-Sh.

14. D(iii) is true (since D is a Sh-Sh and D(ii) is false), ∴ Q not married to a W-W. ∴ Q not married to E. ∴ by elimination Q is married to D and P is married to E. And since D is a Sh-Sh, ∴ Q is a Sh-Sh. ∴ by elimination P is a Pu.

Complete Solution

Alf is a *Pukka* and is married to *Tess* who is a *Wotta-Woppa*.
Bert is a *W-Sh* and is married to *Rachel* who is a *P-Sh*.
Charlie is a *P-Sh* and is married to *Sarah* who is a *W-Sh*.
Duggie is a *Sh-Sh* and is married to *Queenie* who is a *Sh-Sh*.
Ernie is a *W-W* and is married to *Priscilla* who is a *Pu*.
Queenie has five children.

92. More Imperfect Soccer

1. Consider C's figures:

	Played	Won	Lost	Tied	Goals for	Goals against
C	5	2	0	3	3	4

With figures given (W 2, L 0), Goals for would have to be greater than Goals against. ∴ C is either W-W or Sh-Sh. Suppose C Sh-Sh and 1st statement true; then W and T must be false and L true. ∴ W, T must be 1, 4 or 3, 2 to make total 5. In each case Goals for would be greater than Goals against (since L is 0). Goals for (3) is correct and Goals against can only be 3 or 5. In neither case would Goals for be greater than Goals against. ∴ it is not possible for C to be a Sh-Sh, with 1st statement true.

Suppose C Sh-Sh with 1st statement false. Then 'Pl.' must be 4 (not 6 for no team can play more than 5 games) and L must be 1. (W and T remain the same.) This is not possible since W + L + T > Pl. ∴ *C is W-W and all figures are 1 out.* Thus:

Played	Won	Lost	Tied	Goals for	Goals against
4	1	1	2	3 ± 1	4 ± 1

2. Consider A:

	Played	Won	Lost	Tied	Goals for	Goals against
A	3	3	0	0	2	0

This cannot be correct, for it is not possible to win 3 games and only score 2 goals.

Suppose A Sh-Sh with 1st statement true. Figures could only be:

Played	Won	Lost	Tied	Goals for	Goals against
3	2	0	1	2	1

But this is not possible since with 2 W and 0 L, Goals for would have to be at least 2 greater than Goals against.

Suppose A Sh-Sh with 2nd statement true. Figures would be:

Played	Won	Lost	Tied	Goals for	Goals against
4	3	1	0	1 or 3	0

But it is not possible to lose a game with no goals against. ∴ A not Sh-Sh, but W-W. And we have:

	Played	Won	Lost	Tied	Goals for	Goals against
A	4	2	1	1	3	1

(Games against must be 3, for it is not possible to win 2 games, and only score 1 goal.)

3. There are no other W-W's, but 2 Sh-Sh's and 2 P's. Consider D. $W + L + T > 4$ Pl. ∴ D is Sh-Sh. Consider E. With all games tied, Goals for would equal Goals against. But it does not. E is Sh-Sh. ∴ B and F are P's.

4. Consider E:

	Played	Won	Lost	Tied	Goals for	Goals against
E	5	0	0	5	3	2

Suppose 1st statement false, then situation would be:

Played	Won	Lost	Tied
4	0	1	5

which is impossible. ∴ 1st statement true. And we have:

	Played	Won	Lost	Tied	Goals for	Goals against
E	5	1	0	4	3	1

(Goals against is 1 and not 3, because Goals for must be greater than Goals against.)

5. The information we have can be set out in the table thus:

	Played	Won	Lost	Tied	Goals for	Goals against
A	4	2	1	1	3	1
B	3	1	2	0	7	3
C	4	1	1	2	2 or 4	3 or 5
D						
E	5	1	0	4	3	1
F	5	0	4	1	0	11

And we know that D is Sh-Sh.

If D played 4, as given, total of games would be 25. But it must be an even number since each game appears twice. ∴ D's 1st, 3rd and 5th statements are false. ∴ D had no goals against, ∴ D lost no games. ∴ D played 5 games (2nd and 4th statements true). ∴ figures are:

	Played	Won	Lost	Tied	Goals for	Goals against
D	5	3	0	2	2 or 4	0

And since D won 3 matches, Goals for is at least 3 greater than Goals against. ∴ Goals for must be 4. And table is complete except for C's Goals for and Goals against.

6. A table will help to get results of games and scores:

	A	B	C	D	E	F
A	×				T	
B		×			L	
C			×		T	
D				×	T	
E	T	W	T	T	×	T
F					T	×

B tied none, ∴ E's 4 tied games were against A, C, D, F. And E won against B.

(Mark in diagram as shown; other results and scores should be filled in as discovered.)

7. B played 3. D, E and F played everyone. ∴ the 2 sides against which B did not play were A and C.

8. F tied E and lost their other 4 matches. B won against F and lost against E, ∴ lost against D (lost 2). A, B, C, F all tied E; C's and D's other tied game could only be against each other. C's other game (against A) was lost; and A's other game (against D) was lost. *We now know the result of every game.*

9. F scored no goals. ∴ score against E was 0–0, and scores against other sides were 0–?.

D had no Goals against. ∴ scores against C and E were 0–0, and scores against other sides were ?–0.

10. A lost 1 game (against D) and had only 1 Goal against. ∴ A vs. D was 0–1, and A vs. E (tied) was 0–0. A had 1 Goal against (by D), ∴ A vs. C was ?–0.

11. E played 5 and tied 4; totals of goals were 3–1. ∴ E's Won game (against B) was won by a margin of 2 goals (3–1 or 2–0).

B had 3 goals against them by D and E; at least 1 by D (D beat B), ∴ not more than 2 by E. ∴ E vs. B was 2–0. ∴ B vs. D was 0–1. ∴ B vs. F was 7–0.

12. From E's totals, E vs. C (the only game of E's of which we do not know the score) was 1–1. From D's total, D vs. F (the only match of D's of which we do not know the score) was 2–0. ∴ from F's totals F vs. A and F vs. C were both 0–1.

13. From A's totals A vs. C was 2–0.

14. We now know the score in each game. And we know that the totals of C's goals are 2–3.

Complete Solution

(i) Correct table:

	Played	Won	Lost	Tied	Goals for	Goals against
A	4	2	1	1	3	1
B	3	1	2	0	7	3
C	4	1	1	2	2	3
D	5	3	0	2	4	0
E	5	1	0	4	3	1
F	5	0	4	1	0	11

(ii) B and F are Pukkas.
D and E are Shilli-Shallas.
A and C are Wotta-Woppas.

(iii)
	A	B	C	D	E	F
A	×	×	2–0	0–1	0–0	1–0
B	×	×	×	0–1	0–2	7–0
C	0–2	×	×	0–0	1–1	1–0
D	1–0	1–0	0–0	×	0–0	2–0
E	0–0	2–0	1–1	0–0	×	0–0
F	0–1	0–7	0–1	0–2	0–0	×

93. The Age of Remembrance

(i) A table will help:

	P	V	R	F	H
A	×	×	×		×
B		×	×		
C	×	×	✓	×	×
D			×	×	
E	×	✓	×	×	×

Insert in table, as shown, the information that C would marry R and E would marry V. Insert also, as shown, that A would not marry P or H. This enables us to deduce that A would marry F. There is also the prediction that B would not marry H. This enables us to deduce, by elimination, that *D would marry H*. And then, by elimination, that *B would marry P*. This gives us the 5 weddings that would have resulted from the predictions.

(ii) We are told that all the predictions turned out to be incorrect. ∴ none of weddings in (i) took place. Thus:

	P	V	R	F	H
A				×	
B	×				
C			×		
D		×			×
E		×			

Since every prediction is incorrect, A does marry either P or H, ∴ none of the others. And since E predicted that B would not marry H, ∴ *B did marry H*. ∴ *A married P*.

Table now:

	P	V	R	F	H
A	✓	×	×	×	×
B	×	×	×	×	✓
C	×		×		×
D	×				×
E	×	×			×

We are told that if we knew who E's sister was (to whom he cannot be married) we would be able to complete the table. Obviously to enable us to complete the table E's sister must be R or F. *Suppose F.* Then E must be married to R. But we cannot tell which of C and D is married to V and which to F. But we are told that if we knew who E's sister was we would be able to solve the complete problem. ∴ E's sister is not F, but R. ∴ *E must be married to F.* ∴ by elimination *D is married to R, and C to V.*

Complete Solution

(i) Arthur married to Fanny.
    Basil " " Polly.
    Clarence " " Ruth.
    Desmond " " Helen.
    Ethelred " " Veronica.

(ii) Arthur married to Polly.
    Basil " " Helen.
    Clarence " " Veronica.
    Desmond " " Ruth.
    Ethelred " " Fanny.

Ethelred's sister was *Ruth.*

94. Fast or Slow

The facts about Q and S, and P and S can be represented thus:

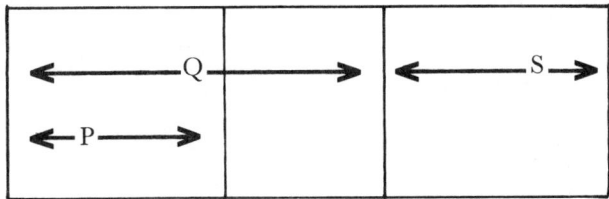

P and R overlap and together cover the whole area. T and S overlap, but do not cover the whole area. Complete situation can therefore be represented thus:

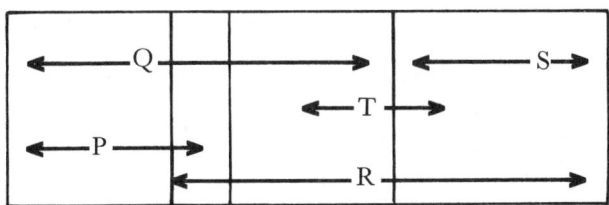

If Smith is a member of P we can see that he must also be a member of Q, and cannot be a member of S. He might or might not be a member of R and/or T.

If Jones is not a member of Q he must be a member of S and R, is not a member of P, and might or might not be a member of T.

If Adam belongs to 4 clubs they must be P, Q, R and T.

A glance at the diagram shows that it is not possible for anyone to belong to only one club.

Complete Solution
- (i) Smith is also a member of the Quicker; he is not a member of the Staiput. We cannot tell whether he is a member of the Rational or the Tortoise.
- (ii) Jones is a member of the Staiput and the Rational. He is not a member of the Progress. We cannot tell whether he is a member of the Tortoise.
- (iii) Progress, Quicker, Rational, Tortoise.
- (iv) No.

95. Help-the-Boys Hall

(i) Since 1 and 2 are contradictories the statement that a certain boy is either in B House or the Green House and the statement that the same boy is either in A House, the White House or Mr. Smith's House, cannot both be true; ∴ the 2 houses mentioned in the first statement are different from the 3 houses mentioned in the second statement; ∴ there are at least 5 houses.

Also since 1 and 2 are contradictories the two statements cannot both be false; ∴ it is not possible for a boy to be in *none* of the houses mentioned in 1 and 2.

∴ *there are only 5 houses.* Call them A, B, C, D, E; their names are Green, White, Yellow, Red and X; their housemasters are Smith, Jones, Budd, Codd and Z.

(ii) The following table will help:

	Green	White	Yellow	Red	X	Mr. Smith	Mr. Jones	Mr. Budd	Mr. Codd	Mr. Z
A	×	×				×				
B	×	×				×				
C				×						
D							×			
E										
Mr. Smith	×	×								
Mr. Jones										
Mr. Budd			×							
Mr. Codd										
Mr. Z										

Mark in the table with crosses (as shown) the information derived from the fact that the houses mentioned in each statement are different: e.g. B House *not* the Green House etc.

Since 1 and 2 are contradictories, B House cannot be the White House or Mr. Smith's; and the Green House cannot be A House or Mr. Smith's.

(The negative information obtained so far has been inserted in diagram. The reader is advised to insert other facts as they are discovered.)

Similar information follows from the fact that 2 and 3 are

also contradictories, and since between them they mention all five houses, therefore the two mentioned in 1 must be the same as the two mentioned in 3. Therefore B House is Mr. Jones's, and D House is the Green House. Insert ticks and fill up rows and columns with crosses.

Notice that since D House is the Green House, and the housemaster of D is not Mr. Jones, therefore Mr. Jones is not the housemaster of the Green House. Similar negative information may be obtained from other positive results.

Since 3 and 4 are contraries four different houses must be referred to. Therefore D House is not Mr. Budd's, etc. Similar deductions can be made from the fact that 4 and 5 are contraries.

Insert facts that Mr. Codd and Mr. Budd are respectively not housemasters of D House and the White House.

By elimination we can now see that Mr. Z must be housemaster of D, and must therefore live in the Green House.

By elimination Mr. Codd lives in the White House.

By elimination Mr. Smith lives in the Yellow House.

By elimination Mr. Jones lives in the Red House and Mr. Budd in X House.

Mr. Jones lives in Red House and is housemaster of B, ∴ B is Red House.

Mr. Smith lives in Yellow House, Yellow House not C, ∴ Mr. Smith not housemaster of C. ∴ by elimination Mr. Smith housemaster of E.

By elimination Mr. Budd housemaster of A, and Mr. Codd of C.

Complete Solution

Letter	Name	Housemaster
A	X	Budd
B	Red	Jones
C	White	Codd
D	Green	Z
E	Yellow	Smith

96. The Light of Truth

Remarks are (avoiding the first person singular):
S. (i) B is heavier than S.
   (ii) B is heavier than J.
J. (i) J is heavier than S.
   (ii) S is heavier than B.
B. (i) S is heavier than B.
   (ii) J weighs the same as B.

Suppose B(ii) true, then since J(ii) and B(i) are the same J(i) must be true (if J and B weigh the same they must have the same number of true remarks). But if J(i) is true, S is more truthful than J, ∴ S(i) and S(ii) must both be true. But S(ii) disagrees with B(ii). ∴ *B(ii) must be false.*

∴ J and B do *not* make the same number of true statements, and since J(ii) and B(i) are the same, ∴ *J(i) must be true.*

∴ J is more truthful (and therefore lighter) than B. ∴ *S(ii) is true.*

Since we know that S has one true and B has one false it is not possible for S to have a *greater* number of false remarks than B.

∴ *J(ii) and B(i) are false.*

And since we know that both B's remarks are false and that S(ii) is true, ∴ *S(i) is true.*

Complete Solution
Brown is the heaviest (2 false remarks);
Jones is next (1 false, 1 true);
Smith is the lightest (2 true remarks).

97. Anstruther and Others

A diagram will be helpful in which facts may be inserted as they are obtained.

	Jo.	Ch.	Law.	De.	Ages
Anstruther					35
Banks			×		
Clopp					40
Dingle		×	×		

(i) B not Law. (his cousin); D not Law. (never been to Scotland); D not Ch. (never driven a car).

(These facts have been inserted in diagram; other facts should be added as they are obtained.)

(ii) If A (35) De., then by elimination, C (40) Law. But Law. is more than 5 years older than De. ∴ A not De.

(iii) If C (40) De., then by elimination A (35) Law. (no one else can be Law.). But Law. older than De., ∴ C not De.

(iv) If C (40) Ch., then by elimination A (35) Law., and B and D between them Jo. and De.

But Jo. would be 43 (3 years older than Ch.), and De. less than 30 (since Law. more than 5 years older than De.). But B and D only differ by 1 year. ∴ C not Ch.

(v) If A (35) Law., then by elimination C (40) Jo., and B and D between them Ch. and De.

Ch. would then be 37 (3 years younger than Jo.), and De. less than 30 (more than 5 years younger than Law.). But B and D only differ by one year, ∴ A not Law.

∴ by elimination C is Law.

(vi) If A (35) Ch. then he is older than De., because Law. C (40) is more than 5 years older than De. And Jo. (3 years older than Ch.) would be 38. But B and D between them are

Jo. and De., and their ages only differ by one year. ∴ A not Ch.

 (vii) ∴ B is Ch. (elimination)
and   D is De. ("   )
and   A is Jo. ("   )

 (viii) Since Ch.(B) is 3 years younger than Jo.(A), who is 35, ∴ B is 32. And since D is a year younger than B, ∴ D is 31.

Complete Solution

Anstruther	Journalist	35
Banks	Chauffeur	32
Clopp	Lawyer	40
Dingle	Dentist	31

98. Abracadabra Avenue

Each person thinks that the numbers of the other two are either above or not above 23 (>23 or $\not> 23$), a perfect square or not a perfect square (sq or $\overline{\text{sq}}$), a multiple of 4 or not a multiple of 4 (m (4) or $\overline{\text{m}}$ (4)).

The combinations of these alternatives with the possibilities in each case are:

    (i) >23, sq, m (4): 36, 64.
    (ii) >23, sq, $\overline{\text{m}}$ (4): 25, 49.
    (iii) >23, $\overline{\text{sq}}$, m (4): 24, 28, 32, 40, 44, etc.
    (iv) >23, $\overline{\text{sq}}$, $\overline{\text{m}}$ (4): 26, 27, 29, 30, etc.
    (v) $\not> 23$, sq, m (4): 4, 16.
    (vi) $\not> 23$, sq, $\overline{\text{m}}$ (4): 1, 9.
    (vii) $\not> 23$, $\overline{\text{sq}}$, m (4): 8, 12, 20.
    (viii) $\not> 23$, $\overline{\text{sq}}$, $\overline{\text{m}}$ (4): 2, 3, 5, 6, 7, etc.

It is important to notice that there is no group which contains only one number, but there are four groups which contain two and one which contains three. Obviously these are the groups in which we are interested.

From what he says A must *either* think that B and C are both in the same group containing three numbers, of which his (A's) is one, or must think that B and C are in the same group containing two numbers.

B thinks he knows A's number, $\therefore$ he must think that it is in a group of two of which his (B's) is the other.

C thinks he knows B's number, $\therefore$ he must think that it is in a group of two of which his (C's) is the other.

Since A thinks B always tells the truth, and C thinks that B always lies, the information which they think they have about B's number is directly contradictory. Pairs of contradictories are (i) and (viii), (ii) and (vii), (iii) and (vi), (iv) and (v).

The only contradictory pair of which one does not contain more than three numbers is (ii) and (vii). And since we know that C must think that B's number is in a group with two possibilities of which his (C's) is the other, C's number must be one of group (ii) (25 or 49) and he must think that B lives in the other one. And A's number must be one of group (vii) (8, 12 or 20) and he must think that B and C live in the other two.

B must also live in a group with two possibilities of which he thinks that A lives in the other. But this group cannot be (ii) ("of the numbers announced, not one is the number of any of the three houses").

∴ B must live in one of the houses of groups (i), (v) or (vi).

We know that one of the numbers announced is eight times the number of one of the houses. Of the numbers that could be announced by A and C (one from group (ii), and the other from (vii)) the only one that is a multiple of 8 is the number 8 which would be announced by A as B's number. If this is eight times the number of one of the houses, it could only be of B's, but it is not posssible for B's or C's number to be 1 as 'their numbers ascend in the order A, B, C.'

∴ the number announced by B must be eight times the number of one of the houses, and we know that the number announced by B must be in group (i), (v), or (vi).

Not (vi) (no multiple of 8); not (v) ((16 is a multiple of 8, but we know that 2 is not the number of any of their houses as it is in group (viii)).

∴ (i). 64 must be eight times the number of A's house) *and B must live in 36.*

And since their numbers ascend in the order A, B, C, C must live in *49* and not 25.

Complete Solution
  A lives in No. 8.
  B lives in No. 36.
  C lives in No. 49.

99. What Tulsa Tortoise

Tables will be helpful, thus:

	P	Q	R	S	T	Alf	Bert	Charlie	Duggie	Ernie
P	×									
Q		×			2–2		×			×
R			×							
S				×						
T		2–2						×	×	×

In the first part of this table the score in each match can be inserted, as it is discovered; in the second part any information can be inserted, positive or negative, about the team to which each of the five men belong.

1. Consider B(i) and E(ii). These cannot both be true; nor can one of them be true and the other false, for in this case the score of each side in the untrue statement would be *two* goals out. ∴ they must both be false, and score must have been 2–2.

Since both the remarks are false, neither B nor E can be a member of either Q or T.

(These facts have been marked in table; other facts should be inserted as found.)

2. Consider A(i) and D(i). By an argument similar to that in (1), these must both be false and correct score in P vs. R must have been 4–3.

∴ neither A nor D can be a member of either P or R.

3. Consider B(ii), C(i), D(ii). These all say the same thing (S vs. Q was 0–0). But they cannot all be true, since B, C and D cannot all be members of Q or S. ∴ they are all false.

∴ neither B, C, nor D are members of either Q or S. And since scores in incorrect remarks are 1 goal out for each side, ∴ score in S vs. Q must have been 1–1.

4. We now know that D is not a member of P, Q, R or S. ∴ D is a member of T. ∴ no one else is a member of T.

5. Since B, C, D and E are not members of Q, ∴ A is a

member of Q. ∴ A not a member of S. ∴ by elimination E is a member of S. ∴ E not a member of any other team.

6. We now know that one of B and C is a member of P and the other is a member of R.

7. Consider E(i). ('Q vs. R was 0–0.') We know E is not Q or R. ∴ this is false. ∴ score was 1–1.

8. ∴ B(iii) is true, ∴ B a member of R. ∴ by elimination C a member of P.

9. ∴ C(ii) true ('P vs. Q was 1–3').

10. Consider E(iii). ('P vs. S was 5–1.') Since E is in S this is true.

11. Consider D(iii). ('T vs. S was 4–0.') Since D is in T this is true.

12. We are told that P scored twice as many goals as Q. Q scored 7 goals, ∴ P scored 14. P scored 10 (1 + 4 + 5) against Q, R, and S; ∴ P scored 4 against T.

13. Consider A(ii). ('R vs. T was 3–0.') We know that A is not in R or T, ∴ this is false. ∴ true score is either 2–1 or 4–1.

14. We now know everything except T's score against P, R's score against T (though we know this is either 2 or 4) and the score in R vs. S.

Let T's score against P = x.
Let R's score against T = y (y = 2 or 4).
Let score in R vs. S = p − q.

Then S's goal average = $\dfrac{1+1+q+0}{5+1+p+4} = \dfrac{q+2}{p+10}$ and

T's goal average = $\dfrac{x+2+1+4}{4+2+y+0} = \dfrac{x+7}{y+6}$.

And we know that $\dfrac{q+2}{p+10} = \dfrac{x+7}{y+6}$.

15. We know y = 2 or 4; suppose 2. Then $\dfrac{x+7}{y+6}$ is at least $\dfrac{7}{8}$. Greatest possible value of $\dfrac{q+2}{p+10}$ (remembering that

no side scored more than five goals) is $\frac{7}{10}$ which is less than $\frac{7}{8}$.

∴ y cannot be 2, and must be 4.

16. We now have: $\frac{q+2}{p+10} = \frac{x+7}{10}$. Greatest possible value of left-hand side is when q = 5 and p = 0; value is then $\frac{7}{10}$. Least possible value of right-hand side is when x = 0; value is then $\frac{7}{10}$.

∴ these are only possible values of p, q and x which will make this equation true. ∴ p = 0, q = 5, x = 0.

Complete Solution

	P	Q	R	S	T	A	B	C	D	E
P	×	1–3	4–3	5–1	4–0	×	×	✓	×	×
Q	3–1	×	1–1	1–1	2–2	✓	×	×	×	×
R	3–4	1–1	×	0–5	4–1	×	✓	×	×	×
S	1–5	1–1	5–0	×	0–4	×	×	×	×	✓
T	0–4	2–2	1–4	4–0	×	×	×	×	✓	×

100. The Wumbling Widgets

It will be convenient to put the data in more compact form, using obvious abbreviations.

1. Lid off, B.B., F.L. Back } → E.B., A.Q.
2. B.B., S.C. off, Tap C } → A.Q., I.R. blue
3. Lid off, Tap C, F.L. Back } → E.B., I.R. blue
4. S.C. off, Tap C, F.L. Back } → I.R. blue
5. E.B., I.R. blue } → W.W.
6. A.Q., I.R. blue } → S.S., Spr. off

It is important to notice that from 'If A then B' we can *not* deduce 'If not A then not B,' but we can deduce 'If not B then not A.' For example we know that running out of gas will cause the engine of a car to stop, but we cannot deduce that if the engine stops it must have run out of gas. There are many other things which cause engines to stop.

Consider the 6 items above. (5) and (6) are the only ones which give direct information about the 3 faults we are investigating. Let us look at them first.

From (6) we deduce that neither A.Q. nor I.R. blue are the cause of W.W., for if they were they would → W.W.

∴ from (5) *E.B. is the cause of W.W.*

From (5) neither E.B. nor 'I.R. blue' are the cause of S.S. or of 'Spr. off.'

∴ from (6) *A.Q. is cause of S.S.* and *A.Q. is cause of 'Spr. off.'*

We now want to find causes of E.B. and A.Q.

From (2) and (4) neither B.B. nor 'S.C. off' nor Tap C nor F.L. Back can be the cause of E.B.

∴ from (1) and (3) *Lid off is cause of E.B.*

Consider A.Q.

From (3) and (4) neither 'Lid off,' Tap C, 'F.L. Back' nor

305

'S.C. off' can be cause of A.Q.
∴ from (1) and (2) *B.B. must be cause of A.Q.*

Complete Solution

To prevent the widgets from wumbling refrain from taking off the lid, which will prevent the engine from boiling.

To prevent the stugs from sticking and the sprockets from falling off refrain from pressing button B; this will prevent the anemometer from quivering.

101. 13th Avenue

It is clear that from Smith's answers to Jones's 3 questions we can get, to start with, no help at all. The starting point must be the statement by Jones: "If I knew whether or not the second figure was 1, I could tell you the number of the house." It will be helpful to consider what Jones must be thinking, even though much of his data and his conclusion are wrong. Jones is in a position where he thinks he has reduced the possible alternatives to two, and one of them has 1 as its second figure.

If Jones were to think that the number was a square but not a cube there would be far too many alternatives (the squares of all numbers from 4 to 22 inclusive lie between 13 and 500, and of all numbers from 23 to 36 inclusive between 500 and 1300). It looks therefore as though he must think that the number is a cube.

The relevant cubes are 27, 64, 125, 216, 343, 512, 729, 1000. (They are the cubes of 3, 4, 5, 6, 7, 8, 9, 10.) Of these 64 and 729 are also squares (of 8 and 27 respectively).

If Jones were to think that the number was a perfect square *and* a perfect cube less than 500 there would be no alternative in his mind — it would *have* to be 64. And if he were to think it a perfect square *and* a perfect cube greater than 500 it would have to be 729. If he were to think it a perfect cube and not a perfect square less than 500 there would be four possibilities (27, 125, 216, 343); but if he were to think it a perfect cube and not a perfect square above 500 there are only two possibilities, 512 and 1000, one of which has 1 as its second figure.

This, then, is what Jones must think.

But he thinks in certain respects incorrectly. He thinks it is not less than 500, but Smith lied to him about this, ∴ it is less than 500.

He thinks it is not a perfect square, but Smith also lied to him about this, ∴ it is a perfect square.

He thinks it is a perfect cube and Smith told him the truth about this, ∴ it is a perfect cube.

∴ the number of Smith's house is a perfect square and a perfect cube less than 500 (and not less than 13), ∴ it can only be *64*.

102. Refined by Mogrification

The sentence that we have to demogrify is:
1. VNRUAOHB
2. VEASNTK
3. UAIO
4. IMENE
5. SKCRTI
6. FO
7. ASNED
8. RHESQIU
9. NTRTSRF
10. TEISFTUL

For ease of reference each 'word' has been numbered.

(a) We note that supermogrification deals only with words that have hitherto been left untouched.

(b) The first step in transmogrification suggests that we should look for a 'word' composed only of consonants and one composed only of vowels. (9) and (3) are the only examples of this. Reverse the consonants and fit the vowels in, and it is easy to see that FRUSTRATION must have been the longest word.

(c) The second step in transmogrification suggests that we look for a word with all the consonants at the beginning and all the vowels at the end. (5) is the only example. This must have been TRICKS and must originally have been the last word.

(d) From the third step in transmogrification we see that (1) and (10) must be the results of bisecting the result of the wedding of two words. Since the first half was reversed, we reverse it again and take a look at (1) reversed and (10).

BHOAURNV and TEISFTUL

It is likely that these are the letters of the two original words re-arranged according to some rule. Try out *simple* rules first. Combine the two and take alternate letters.

We get

BOUNTIFU_ and HARVEST

and it is obvious that the final L must go with the first word.

It looks therefore as though BOUNTIFUL HARVEST

must have been the two words, and the rules for 'wedding' must be:

Take the first letter of the first word and then the first letter of the second and then take letters alternately from the two words in the order in which they come. When the letters of either word are exhausted the remaining letters of the other word shall be taken consecutively.

(e) We have not yet considered (2), (4), (6), (7), (8). It looks as though these must be the results of supermogrification. In particular (6) looks like the middle one of the hitherto untouched words, which has been reversed. It becomes OF. (2) and (8) must clearly be the result of one reverse-wedding and bisection, and (4) and (7) of another. (8) and (7) must be the *first* halves. We look at the pairs:

    RHESQIU and VEASNTK;
and   ASNED and IMENE.

Consider the two words of the first pair together. Alternate letters produce REQUEST and we are left with HSIVANK which is obviously KNAVISH reversed.

The same procedure with the second pair seems at first to produce ANDMN. A little reflection shows, however, that the words we want are probably AND ENEMIES.

The rules for reverse-wedding therefore seem to be: Reverse the second word and then apply the same rules as for a wedding.

(f) Not much thought about the order of the words is now necessary to see that the original message was:

Complete Solution

REQUEST BOUNTIFUL HARVEST AND
FRUSTRATION OF ENEMIES KNAVISH
TRICKS.